Cram101 Textbook Outlines to accompany:

Introduction to Quantum Mechanics

David J. Griffiths, 2nd Edition

A Cram101 Inc. publication (c) 2010.

PRACTICE EXAMS.

Get all of the self-teaching practice exams for each chapter of this textbook at **www.Cram101.com** and ace the tests. Here is an example:

Chapter 1

Introduction to Quantum Mechanics
David J. Griffiths, 2nd Edition,
All Material Written and Prepared by Cram101

I WANT A BETTER GRADE.   Items 1 - 50 of 100.

1 _____ are the expansion coefficients of total angular momentum eigenstates in an uncoupled tensor product basis.

Below, this definition is made precise by defining angular momentum operators, angular momentum eigenstates, and tensor products of these states.

From the formal definition of angular momentum, recursion relations for the _____ can be found.

- ⚪ Clebsch-Gordan coefficients
- ⚪ C parity
- ⚪ C Carinae
- ⚪ Cadmium

2 In mathematics, a _____ is a constant multiplicative factor of a specific object. For example, in the expression $9x^2$, the _____ of x^2 is 9.

The object can be such things as a variable, a vector, a function, etc.

- ⚪ Coefficient
- ⚪ C parity
- ⚪ C Carinae
- ⚪ Cadmium

3 The _____ is the shifting and splitting of spectral lines of atoms and molecules due to the presence of an external static electric field. The amount of splitting and or shifting is called the Stark splitting or Stark shift. In general one distinguishes first- and second-order _____ s.

- ⚪ Stark effect
- ⚪ Saddle point

You get a 50% discount for the online exams. Go to **Cram101.com**, click Sign Up at the top of the screen, and enter DK73DW8135 in the promo code box on the registration screen. Access to Cram101.com is $4.95 per month, cancel at any time.

With Cram101.com online, you also have access to extensive reference material.

You will nail those essays and papers. Here is an example from a Cram101 Biology text:

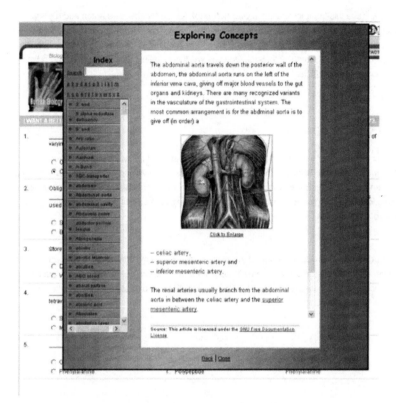

Learning System

Cram101 Textbook Outlines is a learning system. The notes in this book are the highlights of your textbook, you will never have to highlight a book again.

How to use this book. Take this book to class, it is your notebook for the lecture. The notes and highlights on the left hand side of the pages follow the outline and order of the textbook. All you have to do is follow along while your instructor presents the lecture. Circle the items emphasized in class and add other important information on the right side. With Cram101 Textbook Outlines you'll spend less time writing and more time listening. Learning becomes more efficient.

Cram101.com Online

Increase your studying efficiency by using Cram101.com's practice tests and online reference material. It is the perfect complement to Cram101 Textbook Outlines. Use self-teaching matching tests or simulate in-class testing with comprehensive multiple choice tests, or simply use Cram's true and false tests for quick review. Cram101.com even allows you to enter your in-class notes for an integrated studying format combining the textbook notes with your class notes.

Visit **www.Cram101.com**, click Sign Up at the top of the screen, and enter **DK73DW8135** in the promo code box on the registration screen. Access to www.Cram101.com is normally $9.95 per month, but because you have purchased this book, your access fee is only $4.95 per month. Sign up and stop highlighting textbooks forever.

Introduction to Quantum Mechanics
David J. Griffiths, 2nd

CONTENTS

Clebsch-Gordan coefficients	Clebsch-Gordan coefficients are the expansion coefficients of total angular momentum eigenstates in an uncoupled tensor product basis.
	Below, this definition is made precise by defining angular momentum operators, angular momentum eigenstates, and tensor products of these states.
	From the formal definition of angular momentum, recursion relations for the Clebsch-Gordan coefficients can be found.

Coefficient	In mathematics, a coefficient is a constant multiplicative factor of a specific object. For example, in the expression $9x^2$, the coefficient of x^2 is 9.
	The object can be such things as a variable, a vector, a function, etc.

Stark effect	The Stark effect is the shifting and splitting of spectral lines of atoms and molecules due to the presence of an external static electric field. The amount of splitting and or shifting is called the Stark splitting or Stark shift. In general one distinguishes first- and second-order Stark effect s.

Measurement	The framework of quantum mechanics requires a careful definition of measurement and a thorough discussion of its practical and philosophical implications.
	measurement is viewed in different ways in the many interpretations of quantum mechanics; however, despite the considerable philosophical differences, they almost universally agree on the practical question of what results from a routine quantum-physics laboratory measurement. To describe this, a simple framework to use is the Copenhagen interpretation, and it will be implicitly used in this section; the utility of this approach has been verified countless times, and all other interpretations are necessarily constructed so as to give the same quantitative predictions as this in almost every case.

Hamiltonian	In quantum mechanics, the Hamiltonian H is the observable corresponding to the total energy of the system. It is a Hermitian matrix, that, when multiplied by the column vector representing the state of the system, gives a vector representing the total energy of the system. As with all observables, the spectrum of the Hamiltonian is the set of possible outcomes when one measures the total energy of a system.

Mechanics	Mechanics is the branch of physics concerned with the behaviour of physical bodies when subjected to forces or displacements, and the subsequent effect of the bodies on their environment. The discipline has its roots in several ancient civilizations During the early modern period, scientists such as Galileo, Kepler, and especially Newton, laid the foundation for what is now known as classical Mechanics.

Quantum	In physics, a Quantum is an indivisible entity of a quantity that has the same units as the Planck constant and is related to both energy and momentum of elementary particles of matter (called fermions) and of photons and other bosons. The word comes from the Latin 'quantus', for 'how much.' Behind this, one finds the fundamental notion that a physical property may be 'quantized', referred to as 'quantization'. This means that the magnitude can take on only certain discrete numerical values, rather than any value, at least within a range.

Quantum mechanics	Quantum mechanics is a set of principles underlying the most fundamental known description of all physical systems at the submicroscopic scale (at the atomic level.) Notable among these principles are simultaneous wave-like and particle-like behavior of matter and radiation ('Wave-particle duality'), and the prediction of probabilities in situations where classical physics predicts certainties. Classical physics can be derived as a good approximation to quantum physics, typically in circumstances with large numbers of particles.

John Stewart Bell	John Stewart Bell was a physicist, and the originator of Bell's Theorem, one of the most important theorems in quantum physics.
	He was born in Belfast, Northern Ireland, and graduated in experimental physics at the Queen's University of Belfast, in 1948. He went on to complete a PhD at the University of Birmingham, specialising in nuclear physics and quantum field theory.
David Joseph Bohm	David Joseph Bohm was an British quantum physicist who made significant contributions in the fields of theoretical physics, philosophy and neuropsychology, and to the Manhattan Project.
	Bohm was born in Wilkes-Barre, Pennsylvania to a Hungarian Jewish immigrant father and a Lithuanian Jewish mother. He was raised mainly by his father, a furniture store owner and assistant of the local rabbi.
Copenhagen interpretation	The Copenhagen interpretation is an interpretation of quantum mechanics. A key feature of quantum mechanics is that the state of every particle is described by a wavefunction, which is a mathematical representation used to calculate the probability for it to be found in a location or a state of motion. In effect, the act of measurement causes the calculated set of probabilities to 'collapse' to the value defined by the measurement.
Wolfgang Ernst Pauli	Wolfgang Ernst Pauli was an Austrian theoretical physicist noted for his work on spin theory, and for the discovery of the exclusion principle underpinning the structure of matter and the whole of chemistry.
	Pauli was born in Vienna to Wolfgang Joseph Pauli (né Wolf Pascheles) and Berta Camilla Schütz. His middle name was given in honor of his godfather, the physicist Ernst Mach.
Expectation value	In quantum mechanics, the Expectation value is the predicted mean value of the result of an experiment. It is a fundamental concept in all areas of quantum physics.
	Quantum physics shows an inherent statistical behaviour: The measured outcome of an experiment will generally not be the same if the experiment is repeated several times.
Density	The Density of a material is defined as its mass per unit volume. The symbol of Density is ρ '>rho.)
	Mathematically:
	$$\rho = \frac{m}{V}$$
	where:
	ρ is the Density, m is the mass, V is the volume.
Alpha decay	Alpha decay is a type of radioactive decay in which an atomic nucleus emits an alpha particle (two protons and two neutrons bound together into a particle identical to a helium nucleus) and transforms (or 'decays') into an atom with a mass number 4 less and atomic number 2 less. For example:

(The second form is preferred because the first form appears electrically unbalanced. Fundamentally, the recoiling nucleus is very quickly stripped of the two extra electrons which give it an unbalanced charge.

Ensemble

In fluid mechanics, an Ensemble is an imaginary collection of notionally identical experiments.

Each member of the Ensemble will have nominally identical boundary conditions and fluid properties. If the flow is turbulent, the details of the fluid motion will differ from member to member because the experimental setup will be microscopically different; and these slight differences become magnified as time progresses.

Momentum

In classical mechanics, Momentum is the product of the mass and velocity of an object (p = mv.) For more accurate measures of Momentum, see the section 'modern definitions of Momentum' on this page.

Velocity

In physics, Velocity is defined as the rate of change of position. It is a vector physical quantity; both speed and direction are required to define it. In the SI (metric) system, it is measured in meters per second: (m/s) or ms^{-1}.

Energy

In physics, energy is a scalar physical quantity that describes the amount of work that can be performed by a force, an attribute of objects and systems that is subject to a conservation law. Different forms of energy include kinetic, potential, thermal, gravitational, sound, light, elastic, and electromagnetic energy The forms of energy are often named after a related force.

Kinetic energy

The Kinetic energy of an object is the extra energy which it possesses due to its motion. It is defined as the work needed to accelerate a body of a given mass from rest to its current velocity. Having gained this energy during its acceleration, the body maintains this Kinetic energy unless its speed changes.

Boltzmann constant

The Boltzmann constant is the physical constant relating energy at the particle level with temperature observed at the bulk level. It is the gas constant R divided by the Avogadro constant N_A:

$$k = \frac{R}{N_A}.$$

It has the same units as entropy. It is named after the Austrian physicist Ludwig Boltzmann.

Solid

The Solid state of matter is characterized by a distinct structural rigidity and virtual resistance to deformation (i.e. changes of shape and/or volume.) Most Solid s have high values both of Young's modulus and of the shear modulus of elasticity. This contrasts with most liquids or fluids, which have a low shear modulus, and typically exhibit the capacity for macroscopic viscous flow.

Stark effect	The Stark effect is the shifting and splitting of spectral lines of atoms and molecules due to the presence of an external static electric field. The amount of splitting and or shifting is called the Stark splitting or Stark shift. In general one distinguishes first- and second-order Stark effect s.
Hamiltonian	In quantum mechanics, the Hamiltonian H is the observable corresponding to the total energy of the system. It is a Hermitian matrix, that, when multiplied by the column vector representing the state of the system, gives a vector representing the total energy of the system. As with all observables, the spectrum of the Hamiltonian is the set of possible outcomes when one measures the total energy of a system.
Energy	In physics, energy is a scalar physical quantity that describes the amount of work that can be performed by a force, an attribute of objects and systems that is subject to a conservation law. Different forms of energy include kinetic, potential, thermal, gravitational, sound, light, elastic, and electromagnetic energy The forms of energy are often named after a related force.
Boundary	In thermodynamics, a Boundary is a real or imaginary volumetric demarcation region drawn around a thermodynamic system across which quantities such as heat, mass, or work can flow. In short, a thermodynamic Boundary is a division between a system and its surroundings. A Boundary may be adiabatic, isothermal, diathermal, insulating, permeable, or semipermeable.
Boundary conditions	In mathematics, in the field of differential equations, a boundary value problem is a differential equation together with a set of additional restraints, called the boundary conditions A solution to a boundary value problem is a solution to the differential equation which also satisfies the boundary conditions Boundary value problems arise in several branches of physics as any physical differential equation will have them.
Harmonic	In acoustics and telecommunication, a Harmonic of a wave is a component frequency of the signal that is an integer multiple of the fundamental frequency. For example, if the fundamental frequency is f, the Harmonic s have frequencies f, 2f, 3f, 4f, etc. The Harmonic s have the property that they are all periodic at the fundamental frequency, therefore the sum of Harmonic s is also periodic at that frequency.
Harmonic oscillator	In classical mechanics, a Harmonic oscillator is a system which, when displaced from its equilibrium position, experiences a restoring force, F, proportional to the displacement, x according to Hooke's law: $$F = -kx$$ where k is a positive constant. If F is the only force acting on the system, the system is called a simple Harmonic oscillator, and it undergoes simple harmonic motion: sinusoidal oscillations about the equilibrium point, with a constant amplitude and a constant frequency (which does not depend on the amplitude.) If a frictional force (damping) proportional to the velocity is also present, the Harmonic oscillator is described as a damped oscillator.
Excited state	Excitation is an elevation in energy level above an arbitrary baseline energy state. In physics there is a specific technical definition for energy level which is often associated with an atom being excited to an Excited state. In quantum mechanics an Excited state of a system (such as an atom, molecule or nucleus) is any quantum state of the system that has a higher energy than the ground state (that is, more energy than the absolute minimum.)

Clebsch-Gordan coefficients	Clebsch-Gordan coefficients are the expansion coefficients of total angular momentum eigenstates in an uncoupled tensor product basis.
	Below, this definition is made precise by defining angular momentum operators, angular momentum eigenstates, and tensor products of these states.
	From the formal definition of angular momentum, recursion relations for the Clebsch-Gordan coefficients can be found.
Coefficient	In mathematics, a coefficient is a constant multiplicative factor of a specific object. For example, in the expression $9x^2$, the coefficient of x^2 is 9.
	The object can be such things as a variable, a vector, a function, etc.
Node	In electrical engineering, Node refers to any point on a circuit where two or more circuit elements meet. For two nodes to be different, their voltages must be different. Without any further knowledge, it is easy to establish how to find a Node by using Ohm's Law: V=IR. When looking at circuit schematics, ideal wires have a resistance of zero (this is not true in real life, but it is a good assumption.)
Measurement	The framework of quantum mechanics requires a careful definition of measurement and a thorough discussion of its practical and philosophical implications.
	measurement is viewed in different ways in the many interpretations of quantum mechanics; however, despite the considerable philosophical differences, they almost universally agree on the practical question of what results from a routine quantum-physics laboratory measurement. To describe this, a simple framework to use is the Copenhagen interpretation, and it will be implicitly used in this section; the utility of this approach has been verified countless times, and all other interpretations are necessarily constructed so as to give the same quantitative predictions as this in almost every case.
Angle	In geometry and trigonometry, an Angle Where there is no possibility of confusion, the term Angle is used interchangeably for both the geometric configuration itself and for its angular magnitude (which is simply a numerical quantity.)
Taylor series	In mathematics, the Taylor series is a representation of a function as an infinite sum of terms calculated from the values of its derivatives at a single point. It may be regarded as the limit of the Taylor polynomials. Taylor series are named after the English mathematician Brook Taylor.
Ladder operators	In linear algebra (and its application to quantum mechanics), a raising or lowering operator (collectively known as Ladder operators) is an operator that increases or decreases the eigenvalue of another operator. In quantum mechanics, the raising operator is sometimes called the creation operator, and the lowering operator the annihilation operator. Well-known applications of Ladder operators in quantum mechanics are in the formalisms of the quantum harmonic oscillator and angular momentum.
Mechanics	Mechanics is the branch of physics concerned with the behaviour of physical bodies when subjected to forces or displacements, and the subsequent effect of the bodies on their environment. The discipline has its roots in several ancient civilizations During the early modern period, scientists such as Galileo, Kepler, and especially Newton, laid the foundation for what is now known as classical Mechanics.

Momentum	In classical mechanics, Momentum is the product of the mass and velocity of an object ($p = mv$.) For more accurate measures of Momentum, see the section 'modern definitions of Momentum' on this page.
Quantum	In physics, a Quantum is an indivisible entity of a quantity that has the same units as the Planck constant and is related to both energy and momentum of elementary particles of matter (called fermions) and of photons and other bosons. The word comes from the Latin 'quantus', for 'how much.' Behind this, one finds the fundamental notion that a physical property may be 'quantized', referred to as 'quantization'. This means that the magnitude can take on only certain discrete numerical values, rather than any value, at least within a range.
Quantum mechanics	Quantum mechanics is a set of principles underlying the most fundamental known description of all physical systems at the submicroscopic scale (at the atomic level.) Notable among these principles are simultaneous wave-like and particle-like behavior of matter and radiation ('Wave-particle duality'), and the prediction of probabilities in situations where classical physics predicts certainties. Classical physics can be derived as a good approximation to quantum physics, typically in circumstances with large numbers of particles.
Canonical commutation relation	In physics, the Canonical commutation relation is the relation between canonical conjugate quantities (quantities which are related by definition such that one is the Fourier transform of another), for example: $[x, p_x] = i$ between the position x and momentum p_x in the x direction of a point particle in one dimension, where $[x, p_x] = x p_x - p_x x$ is the so-called commutator of x and p_x, i is the imaginary unit, and Ä§ is the reduced Planck's constant $h/2\pi$. This relation is attributed to Max Born, and it was noted by E. Kennard (1927) to imply the Heisenberg uncertainty principle. By contrast, in classical physics, all observables commute and the commutator would be zero.
Force	In physics, a Force is a push or pull that can cause an object with mass to change its velocity. Force has both magnitude and direction, making it a vector quantity. Newton's second law states that an object with a constant mass will accelerate in proportion to the net Force acting upon and in inverse proportion to its mass.
Frequency	Frequency is the number of occurrences of a repeating event per unit time. It is also referred to as temporal Frequency. The period is the duration of one cycle in a repeating event, so the period is the reciprocal of the Frequency.
Alpha decay	Alpha decay is a type of radioactive decay in which an atomic nucleus emits an alpha particle (two protons and two neutrons bound together into a particle identical to a helium nucleus) and transforms (or 'decays') into an atom with a mass number 4 less and atomic number 2 less. For example: (The second form is preferred because the first form appears electrically unbalanced. Fundamentally, the recoiling nucleus is very quickly stripped of the two extra electrons which give it an unbalanced charge.
Free particle	In physics, a Free particle is a particle that, in some sense, is not bound. In classical physics, this means the particle is present in a 'field-free' space. The classical Free particle is characterized simply by a fixed velocity.
Generating function	In mathematics, a Generating function is a formal power series whose coefficients encode information about a sequence a_n that is indexed by the natural numbers.

There are various types of Generating function s, including ordinary Generating function s, exponential Generating function s, Lambert series, Bell series, and Dirichlet series; definitions and examples are given below. Every sequence has a Generating function of each type.

Velocity	In physics, Velocity is defined as the rate of change of position. It is a vector physical quantity; both speed and direction are required to define it. In the SI (metric) system, it is measured in meters per second: (m/s) or ms^{-1}.
Group velocity	The Group velocity of a wave is the velocity with which the overall shape of the wave's amplitudes -- known as the modulation or envelope of the wave -- propagates through space.
	For example, imagine what happens if a stone is thrown into the middle of a very still pond. When the stone hits the surface of the water, a circular pattern of waves appears.
Dispersion	In fluid dynamics, Dispersion of water waves generally refers to frequency Dispersion. Frequency Dispersion means that waves of different wavelengths travel at different phase speeds. Water waves, in this context, are waves propagating on the water surface, and forced by gravity and surface tension.
Bound state	In physics, a Bound state is a composite of two or more building blocks (particles or bodies) that behaves as a single object. In quantum mechanics (where the number of particles is conserved), a Bound state is a state in the Hilbert space that corresponds to two or more particles whose interaction energy is negative, and therefore these particles cannot be separated unless energy is spent. The energy spectrum of a Bound state is discrete, unlike the continuous spectrum of isolated particles.
Scattering	Scattering is a general physical process where some forms of radiation, such as light, sound are forced to deviate from a straight trajectory by one or more localized non-uniformities in the medium through which they pass. In conventional use, this also includes deviation of reflected radiation from the angle predicted by the law of reflection. Reflections that undergo Scattering are often called diffuse reflections and unscattered reflections are called specular (mirror-like) reflections.
Reflection coefficient	The Reflection coefficient is used in physics and electrical engineering when wave propagation in a medium containing discontinuities is considered. A Reflection coefficient describes either the amplitude or the intensity of a reflected wave relative to an incident wave. The Reflection coefficient is closely related to the transmission coefficient.
Transmission	Using the principle of mechanical advantage, a Transmission or gearbox provides a speed-torque conversion (commonly known as 'gear reduction' or 'speed reduction') from a higher speed motor to a slower but more forceful output or vice-versa. Main gearbox of the Bristol 171 Sycamore helicopter
	Early transmissions included the right-angle drives and other gearing in windmills, horse-powered devices, and steam engines, in support of pumping, milling, and hoisting.
	Most modern gearboxes are used to increase torque while reducing the speed of a prime mover output shaft (e.g. a motor drive shaft.)
Satellite Test of the Equivalence Principle	The Satellite Test of the Equivalence Principle is a proposed (as of 2008) space science experiment to test the equivalence principle of general relativity. The experiment is thought to be sensitive enough to test Einstein's theory of gravity and other theories.

The basic configuration is that of a drag-free satellite where an outer shell around an inner test mass is used to block solar wind, atmospheric drag, the Earth's magnetic field and other effects which might disturb the motion of a freely-falling inner object.

Balmer series

The Balmer series or Balmer lines in atomic physics, is the designation of one of a set of six different named series describing the spectral line emissions of the hydrogen atom. The Balmer series is calculated using the Balmer formula, an empirical equation discovered by Johann Balmer in 1885.

The visible spectrum of light from hydrogen displays four wavelengths, 410 nm, 434 nm, 486 nm, and 656 nm, that reflect emissions of photons by electrons in excited states transitioning to the quantum level described by the principal quantum number n equals 2.

Degenerate

In physics two or more different physical states are said to be degenerate if they are all at the same energy level. Physical states differ if and only if they are linearly independent. An energy level is said to be degenerate if it contains two or more different states.

Hilbert space	The mathematical concept of a Hilbert space generalizes the notion of Euclidean space. It extends the methods of vector algebra from the two-dimensional plane and three-dimensional space to infinite-dimensional spaces. In more formal terms, a Hilbert space is an inner product space -- an abstract vector space in which distances and angles can be measured -- which is 'complete', meaning that if a sequence of vectors is Cauchy, then it converges to some limit within the space.
Clebsch-Gordan coefficients	Clebsch-Gordan coefficients are the expansion coefficients of total angular momentum eigenstates in an uncoupled tensor product basis. Below, this definition is made precise by defining angular momentum operators, angular momentum eigenstates, and tensor products of these states. From the formal definition of angular momentum, recursion relations for the Clebsch-Gordan coefficients can be found.
Coefficient	In mathematics, a coefficient is a constant multiplicative factor of a specific object. For example, in the expression $9x^2$, the coefficient of x^2 is 9. The object can be such things as a variable, a vector, a function, etc.
Hamiltonian	In quantum mechanics, the Hamiltonian H is the observable corresponding to the total energy of the system. It is a Hermitian matrix, that, when multiplied by the column vector representing the state of the system, gives a vector representing the total energy of the system. As with all observables, the spectrum of the Hamiltonian is the set of possible outcomes when one measures the total energy of a system.
Observable	In physics, particularly in quantum physics, a system Observable is a property of the system state that can be determined by some sequence of physical operations. For example, these operations might involve submitting the system to various electromagnetic fields and eventually reading a value off some gauge. In systems governed by classical mechanics, any experimentally Observable value can be shown to be given by a real-valued function on the set of all possible system states.
Balmer series	The Balmer series or Balmer lines in atomic physics, is the designation of one of a set of six different named series describing the spectral line emissions of the hydrogen atom. The Balmer series is calculated using the Balmer formula, an empirical equation discovered by Johann Balmer in 1885. The visible spectrum of light from hydrogen displays four wavelengths, 410 nm, 434 nm, 486 nm, and 656 nm, that reflect emissions of photons by electrons in excited states transitioning to the quantum level described by the principal quantum number n equals 2.
Degenerate	In physics two or more different physical states are said to be degenerate if they are all at the same energy level. Physical states differ if and only if they are linearly independent. An energy level is said to be degenerate if it contains two or more different states.
Momentum	In classical mechanics, Momentum is the product of the mass and velocity of an object (p = mv.) For more accurate measures of Momentum, see the section 'modern definitions of Momentum' on this page.
Rigged Hilbert space	In mathematics, a Rigged Hilbert space is a construction designed to link the distribution and square-integrable aspects of functional analysis. Such spaces were introduced to study spectral theory in the broad sense. They can bring together the 'bound state' (eigenvector) and 'continuous spectrum', in one place.

Stark effect	The Stark effect is the shifting and splitting of spectral lines of atoms and molecules due to the presence of an external static electric field. The amount of splitting and or shifting is called the Stark splitting or Stark shift. In general one distinguishes first- and second-order Stark effect s.
Alpha decay	Alpha decay is a type of radioactive decay in which an atomic nucleus emits an alpha particle (two protons and two neutrons bound together into a particle identical to a helium nucleus) and transforms (or 'decays') into an atom with a mass number 4 less and atomic number 2 less. For example:
	(The second form is preferred because the first form appears electrically unbalanced. Fundamentally, the recoiling nucleus is very quickly stripped of the two extra electrons which give it an unbalanced charge.
Momentum space	The Momentum space associated with a particle is a vector space in which every point $\{k_x, k_y, k_z\}$ corresponds to a possible value of the momentum vector $\vec{k}$. Representing a problem in terms of the momenta of the particles involved, rather than in terms of their positions, can greatly simplify some problems in physics.
	In quantum physics, a particle is described by a quantum state.
Free particle	In physics, a Free particle is a particle that, in some sense, is not bound. In classical physics, this means the particle is present in a 'field-free' space.
	The classical Free particle is characterized simply by a fixed velocity.
Energy	In physics, energy is a scalar physical quantity that describes the amount of work that can be performed by a force, an attribute of objects and systems that is subject to a conservation law. Different forms of energy include kinetic, potential, thermal, gravitational, sound, light, elastic, and electromagnetic energy The forms of energy are often named after a related force.
Neutrinos	Neutrinos are elementary particles that often travel close to the speed of light, lack an electric charge, are able to pass through ordinary matter almost undisturbed and are thus extremely difficult to detect. Neutrinos have a minuscule, but nonzero mass. They are usually denoted by the Greek letter v ' href='/wiki/Nu_(letter)'>nu.)
Angle	In geometry and trigonometry, an Angle Where there is no possibility of confusion, the term Angle is used interchangeably for both the geometric configuration itself and for its angular magnitude (which is simply a numerical quantity.)
Coherent state	In quantum mechanics a Coherent state is a specific kind of quantum state of the quantum harmonic oscillator whose dynamics most closely resemble the oscillating behaviour of a classical harmonic oscillator system. It was the first example of quantum dynamics when Erwin Schrödinger derived it in 1926 while searching for solutions of the Schrödinger equation that satisfy the correspondence principle. The quantum harmonic oscillator and hence, the Coherent state, arise in the quantum theory of a wide range of physical systems.
Displacement	In fluid mechanics, Displacement occurs when an object is immersed in a fluid, pushing it out of the way and taking its place. The volume of the fluid displaced can then be measured, as in the illustration, and from this the volume of the immersed object can be deduced (the volume of the immersed object will be exactly equal to the volume of the displaced fluid.)
	An object that sinks displaces an amount of fluid equal to the object's volume.

Translation

In physics, Translation is movement that changes the position of an object, as opposed to rotation. For example, according to Whittaker:

A Translation is the operation changing the positions of all points (x, y, z) of an object according to the formula

$$(x, y, z) \rightarrow (x + \Delta x, y + \Delta y, z + \Delta z)$$

where $(\Delta x, \Delta y, \Delta z)$ is the same vector for each point of the object. The Translation vector $(\Delta x, \Delta y, \Delta z)$ common to all points of the object describes a particular type of displacement of the object, usually called a linear displacement to distinguish it from displacements involving rotation, called angular displacements.

Heisenberg picture

In physics, the Heisenberg picture is that formulation of quantum mechanics where the operators (observables and others) are time-dependent and the state vectors are time-independent. It stands in contrast to the Schrödinger picture in which operators are constant and the states evolve in time. The two pictures only differ by a time-dependent basis change.

Clebsch-Gordan coefficients	Clebsch-Gordan coefficients are the expansion coefficients of total angular momentum eigenstates in an uncoupled tensor product basis.
	Below, this definition is made precise by defining angular momentum operators, angular momentum eigenstates, and tensor products of these states.
	From the formal definition of angular momentum, recursion relations for the Clebsch-Gordan coefficients can be found.
Canonical commutation relation	In physics, the Canonical commutation relation is the relation between canonical conjugate quantities (quantities which are related by definition such that one is the Fourier transform of another), for example:
	$[x, p_x] = i$
	between the position x and momentum p_x in the x direction of a point particle in one dimension, where $[x, p_x] = x p_x - p_x x$ is the so-called commutator of x and p_x, i is the imaginary unit, and Ä§ is the reduced Planck's constant h /2π . This relation is attributed to Max Born, and it was noted by E. Kennard (1927) to imply the Heisenberg uncertainty principle.
	By contrast, in classical physics, all observables commute and the commutator would be zero.
Coefficient	In mathematics, a coefficient is a constant multiplicative factor of a specific object. For example, in the expression $9x^2$, the coefficient of x^2 is 9.
	The object can be such things as a variable, a vector, a function, etc.
Frequency	Frequency is the number of occurrences of a repeating event per unit time. It is also referred to as temporal Frequency.
	The period is the duration of one cycle in a repeating event, so the period is the reciprocal of the Frequency.
Angle	In geometry and trigonometry, an Angle Where there is no possibility of confusion, the term Angle is used interchangeably for both the geometric configuration itself and for its angular magnitude (which is simply a numerical quantity.)
Associated Legendre function	In mathematics, the Associated Legendre function s are the canonical solutions of the general Legendre equation

$$(1 - x^2)\, y'' - 2xy' + \left(\ell[\ell+1] - \frac{m^2}{1-x^2}\right) y = 0,$$

or

$$([1 - x^2]\, y')' + \left(\ell[\ell+1] - \frac{m^2}{1-x^2}\right) y = 0,$$

where the indices ℓ and m (which in general are complex quantities) are referred to as the degree and order of the Associated Legendre function respectively. This equation has solutions that are nonsingular on [−1, 1] only if ℓ and m are integers with $0 \le m \le \ell$, or with trivially equivalent negative values. When in addition m is even, the function is a polynomial.

Chapter 4. QUANTUM MECHANICS IN THREE DIMENSIONS

Lyman series	In physics and chemistry, the Lyman series is the series of transitions and resulting ultraviolet emission lines of the hydrogen atom as an electron goes from $n \geq 2$ to $n = 1$ (where n is the principal quantum number referring to the energy level of the electron.) The transitions are named sequentially by Greek letters: from $n = 2$ to $n = 1$ is called Lyman-alpha, 3 to 1 is Lyman-beta, 4 to 1 is Lyman-gamma, etc. The series is named after its discoverer, Theodore Lyman.
Azimuthal quantum number	The Azimuthal quantum number symbolized as â„" (lower-case L) is a quantum number for an atomic orbital that determines its orbital angular momentum. The Azimuthal quantum number is the second of a set of quantum numbers (the principal quantum number, following spectroscopic notation, the Azimuthal quantum number, the magnetic quantum number, and the spin quantum number) which describe the unique quantum state of an electron and is designated by the letter â„".
	There is a set of quantum numbers associated with the energy states of the electrons of an atom.
Harmonic	In acoustics and telecommunication, a Harmonic of a wave is a component frequency of the signal that is an integer multiple of the fundamental frequency. For example, if the fundamental frequency is f, the Harmonic s have frequencies f, 2f, 3f, 4f, etc. The Harmonic s have the property that they are all periodic at the fundamental frequency, therefore the sum of Harmonic s is also periodic at that frequency.
Magnetic quantum number	In atomic physics, the Magnetic quantum number is the third of a set of quantum numbers (the principal quantum number, the azimuthal quantum number, the Magnetic quantum number, and the spin quantum number) which describe the unique quantum state of an electron and is designated by the letter m. The Magnetic quantum number denotes the energy levels available within a subshell.
	There are a set of quantum numbers associated with the energy states of the atom.
Quantum	In physics, a Quantum is an indivisible entity of a quantity that has the same units as the Planck constant and is related to both energy and momentum of elementary particles of matter (called fermions) and of photons and other bosons. The word comes from the Latin 'quantus', for 'how much.' Behind this, one finds the fundamental notion that a physical property may be 'quantized', referred to as 'quantization'. This means that the magnitude can take on only certain discrete numerical values, rather than any value, at least within a range.
Effective potential	The Effective potential is a mathematical expression combining angular momentum into the potential energy of a dynamical system. Commonly used in calculating the orbits of planets , the Effective potential allows one to reduce a problem to fewer dimensions.
	One way of thinking of this concept is in terms of the minimum kinetic energy required for an object to escape a gravitational field.
Hamiltonian	In quantum mechanics, the Hamiltonian H is the observable corresponding to the total energy of the system. It is a Hermitian matrix, that, when multiplied by the column vector representing the state of the system, gives a vector representing the total energy of the system. As with all observables, the spectrum of the Hamiltonian is the set of possible outcomes when one measures the total energy of a system.

Atom	The Atom is a basic unit of matter consisting of a dense, central nucleus surrounded by a cloud of negatively charged electrons. The Atom ic nucleus contains a mix of positively charged protons and electrically neutral neutrons (except in the case of hydrogen-1, which is the only stable nuclide with no neutron.) The electrons of an Atom are bound to the nucleus by the electromagnetic force.
Hydrogen	Hydrogen is the chemical element with atomic number 1. It is represented by the symbol H. At standard temperature and pressure, Hydrogen is a colorless, odorless, nonmetallic, tasteless, highly flammable diatomic gas with the molecular formula H_2. With an atomic weight of 1.007 94 u, Hydrogen is the lightest element.
Hydrogen atom	A Hydrogen atom is an atom of the chemical element hydrogen, and an example of a Boson. The electrically neutral atom contains a single positively-charged proton and a single negatively-charged electron bound to the nucleus by the Coulomb force. The most abundant isotope, hydrogen-1, protium, or light hydrogen, contains no neutrons; other isotopes contain one or more neutrons.
Principal quantum number	In atomic physics, the Principal quantum number symbolized as n is the first of a set of quantum numbers (which includes: the Principal quantum number, the azimuthal quantum number, the magnetic quantum number, and the spin quantum number) of an atomic orbital. There are a set of quantum numbers associated with the energy states of the atom. The four quantum numbers n, â„", m, and s specify the complete and unique quantum state of a single electron in an atom called its wavefunction or orbital.
Bohr radius	In the Bohr model of the structure of an atom, put forward by Niels Bohr in 1913, electrons orbit a central nucleus. The model says that the electrons orbit only at certain distances from the nucleus, depending on their energy. In the simplest atom, hydrogen, a single electron orbits, and the smallest possible orbit for the electron, that with the lowest energy, is most likely to be found at a distance from the nucleus called the Bohr radius.
Binding energy	Binding energy is the mechanical energy required to disassemble a whole into separate parts. A bound system has typically a lower potential energy than its constituent parts; this is what keeps the system together. The usual convention is that this corresponds to a positive Binding energy.
Ionization	Ionization is the physical process of converting an atom or molecule into an ion by adding or removing charged particles such as electrons or other ions. This is often confused with dissociation (chemistry.) The process works slightly differently depending on whether an ion with a positive or a negative electric charge is being produced.
Stark effect	The Stark effect is the shifting and splitting of spectral lines of atoms and molecules due to the presence of an external static electric field. The amount of splitting and or shifting is called the Stark splitting or Stark shift. In general one distinguishes first- and second-order Stark effect s.
Density	The Density of a material is defined as its mass per unit volume. The symbol of Density is ρ '>rho.)

Mathematically:

$$\rho = \frac{m}{V}$$

where:

ρ is the Density,
m is the mass,
V is the volume.

Spherical harmonics

In mathematics, the Spherical harmonics are the angular portion of a set of solutions to Laplace's equation. Represented in a system of spherical coordinates, Laplace's Spherical harmonics Y_ℓ^m are a specific set of Spherical harmonics which forms an orthogonal system, first introduced by Pierre Simon de Laplace. Spherical harmonics are important in many theoretical and practical applications, particularly in the computation of atomic orbital electron configurations, representation of gravitational fields, geoids, and the magnetic fields of planetary bodies and stars, and characterization of the cosmic microwave background radiation.

Photon

In physics, a Photon is an elementary particle, the quantum of the electromagnetic field and the basic 'unit' of light and all other forms of electromagnetic radiation. It is also the force carrier for the electromagnetic force. The effects of this force are easily observable at both the microscopic and macroscopic level, because the Photon has no rest mass; this allows for interactions at long distances.

Balmer series

The Balmer series or Balmer lines in atomic physics, is the designation of one of a set of six different named series describing the spectral line emissions of the hydrogen atom. The Balmer series is calculated using the Balmer formula, an empirical equation discovered by Johann Balmer in 1885.

The visible spectrum of light from hydrogen displays four wavelengths, 410 nm, 434 nm, 486 nm, and 656 nm, that reflect emissions of photons by electrons in excited states transitioning to the quantum level described by the principal quantum number n equals 2.

Rydberg constant

The Rydberg constant is a physical constant relating to atomic spectra in the science of spectroscopy. Rydberg initially determined its value empirically from spectroscopy, but it was later found that its value could be calculated from more fundamental constants by using quantum mechanics.

The Rydberg constant represents the limiting value of the highest wavenumber of any photon that can be emitted from the hydrogen atom, or, alternatively, the wavenumber of the lowest-energy photon capable of ionizing the hydrogen atom from its ground state.

Rydberg formula

The Rydberg formula is used in atomic physics to describe the wavelengths of spectral lines of many chemical elements. The formula was invented by the Swedish physicist Johannes Rydberg and presented on November 5, 1888.

In the 1880s, Rydberg worked on a formula describing the relation between the wavelengths in spectral lines of alkali metals.

Angular momentum	In physics, Angular momentum is a measure of the motion of mass about a center of rotation; by analogy with momentum, the product of inertia (mass) and velocity, Angular momentum can be interpreted as the product of a rotational inertia (moment of inertia or angular mass) and angular velocity. In two dimensions, the Angular momentum of mass m with respect to a chosen origin is given by: $$L = mvr \sin\theta$$ where m is the mass, v is the speed (magnitude of the velocity vector), r is the distance from the origin and θ is the angle between the velocity and the radius vector. This equation can be interpreted as the product of the moment of inertia mr^2 times angular velocity vsin(θ) / r.
Force	In physics, a Force is a push or pull that can cause an object with mass to change its velocity. Force has both magnitude and direction, making it a vector quantity. Newton's second law states that an object with a constant mass will accelerate in proportion to the net Force acting upon and in inverse proportion to its mass.
Ladder operators	In linear algebra (and its application to quantum mechanics), a raising or lowering operator (collectively known as Ladder operators) is an operator that increases or decreases the eigenvalue of another operator. In quantum mechanics, the raising operator is sometimes called the creation operator, and the lowering operator the annihilation operator. Well-known applications of Ladder operators in quantum mechanics are in the formalisms of the quantum harmonic oscillator and angular momentum.
Rigid rotor	The Rigid rotor is a mechanical model that is used to explain rotating systems. An arbitrary Rigid rotor is a 3-dimensional rigid object, such as a top. To orient such an object in space three angles are required.
Spin	In quantum mechanics, Spin is a fundamental property of atomic nuclei, hadrons, and elementary particles. For particles with non-zero Spin, Spin direction (also called Spin for short) is an important intrinsic degree of freedom. The head-on collision of a quark (the red ball) from one proton (the orange ball) with a gluon (the green ball) from another proton with opposite Spin, Spin is represented by the blue arrows circling the protons and the quark.
Classical electron radius	The Classical electron radius is based on a classical relativistic model of the electron. Its value is calculated as $$r_{\mathrm{e}} = \frac{1}{4\pi\epsilon_0}\frac{e^2}{m_e c^2} = 2.8179402894(58) \times 10^{-15}\mathrm{m}$$ where e and m_e are the electric charge and the mass of the electron, c is the speed of light, and ϵ_0 is the permittivity of free space. In cgs units, this becomes more simply $$r_{\mathrm{e}} = \frac{e^2}{m_e c^2} = 2.8179402894(58) \times 10^{-13}\mathrm{cm}$$

with

$$e = 4.80 \times 10^{-10}\text{esu}, m = 9.11 \times 10^{-28}\text{g}, c = 3.00 \times 10^{10}\text{cm/sec}$$

Electron

The Electron is a subatomic particle that carries a negative electric charge. It has no known substructure and is believed to be a point particle. An Electron has a mass that is approximately 1836 times less than that of the proton.

Spinors

In mathematics and physics, in particular in the theory of the orthogonal groups (such as the rotation or the Lorentz groups), Spinors are elements of a complex vector space introduced to expand the notion of spatial vector. They are needed because the full structure of the group of rotations in a given number of dimensions requires some extra number of dimensions to exhibit it. Specifically, Spinors are geometrical objects constructed from a vector space endowed with a quadratic form, such as a Euclidean or Minkowski space, by means of an algebraic procedure, through Clifford algebras, or a quantization procedure.

Eigenspinors

In quantum mechanics, Eigenspinors are thought of as basis vectors representing the general spin state of a particle. Strictly speaking, they are not vectors at all, but in fact spinors. For a single spin 1/2 particle, they can be defined as the eigenvectors of the Pauli matrices.

Gyromagnetic ratio

In physics, the Gyromagnetic ratio of a particle or system is the ratio of its magnetic dipole moment to its angular momentum, and it is often denoted by the symbol γ, gamma. Its SI units are radian per second per tesla (s^{-1}·T^{-1}) or, equivalently, coulomb per kilogram (C·kg^{-1}.)

The term 'Gyromagnetic ratio' is sometimes used as a synonym for a different but closely-related quantity, the g-factor.

Larmor precession

In physics, Larmor precession is the precession of the magnetic moments of electrons, atomic nuclei, and atoms about an external magnetic field. The magnetic field exerts a torque on the magnetic moment,

$$\vec{\Gamma} = \vec{\mu} \times \vec{B} = \gamma \vec{J} \times \vec{B}$$

where $\vec{\Gamma}$ is the torque, $\vec{J}$ is the angular momentum vector, $\vec{B}$ is the external magnetic field, $\times$ is the cross product, and γ is the gyromagnetic ratio which gives the proportionality constant between the magnetic moment and the angular momentum. The angular momentum vector $\vec{J}$ precesses about the external field axis with an angular frequency known as the Larmor frequency,

ω = γB

$$\gamma = \frac{-eg}{2m}$$

where ω is the angular frequency, $\frac{-eg}{2m}$ is the gyromagnetic ratio, and B is the magnitude of the magnetic field and g is the g-factor (normally 1, except for in quantum physics.)

Singlet

In theoretical physics, a singlet usually refers to a one-dimensional representation (e.g. a particle with vanishing spin.) It may also refer to two or more particles prepared in a correlated state, such that the total angular momentum of the state is zero.

	Singlets frequently occur in atomic physics as one of the two ways in which the spin of two electrons can be combined; the other being a triplet.
Harmonic oscillator	In classical mechanics, a Harmonic oscillator is a system which, when displaced from its equilibrium position, experiences a restoring force, F, proportional to the displacement, x according to Hooke's law: $$F = -kx$$ where k is a positive constant. If F is the only force acting on the system, the system is called a simple Harmonic oscillator, and it undergoes simple harmonic motion: sinusoidal oscillations about the equilibrium point, with a constant amplitude and a constant frequency (which does not depend on the amplitude.) If a frictional force (damping) proportional to the velocity is also present, the Harmonic oscillator is described as a damped oscillator.
Continuity equation	A Continuity equation in physics is a differential equation that describes the transport of some kind of conserved quantity. Since mass, energy, momentum, electric charge and other natural quantities are conserved, a vast variety of physics may be described with Continuity equation s. Continuity equation s are the (stronger) local form of conservation laws.
Momentum space	The Momentum space associated with a particle is a vector space in which every point $\{k_x, k_y, k_z\}$ corresponds to a possible value of the momentum vector $\vec{k}$. Representing a problem in terms of the momenta of the particles involved, rather than in terms of their positions, can greatly simplify some problems in physics. In quantum physics, a particle is described by a quantum state.
Alpha decay	Alpha decay is a type of radioactive decay in which an atomic nucleus emits an alpha particle (two protons and two neutrons bound together into a particle identical to a helium nucleus) and transforms (or 'decays') into an atom with a mass number 4 less and atomic number 2 less. For example: (The second form is preferred because the first form appears electrically unbalanced. Fundamentally, the recoiling nucleus is very quickly stripped of the two extra electrons which give it an unbalanced charge.
Rotation	A Rotation is a movement of an object in a circular motion. A two-dimensional object rotates around a center (or point) of Rotation. A three-dimensional object rotates around a line called an axis.
Electromagnetic field	The Electromagnetic field is a physical field produced by electrically charged objects. It affects the behavior of charged objects in the vicinity of the field. The Electromagnetic field extends indefinitely throughout space and describes the electromagnetic interaction.
Motion	In physics, Motion means a change in the location of a body. Change in Motion is the result of applied force. Motion is typically described in terms of velocity, acceleration, displacement, and time.

Chapter 5. IDENTICAL PARTICLES

Identical particles	Identical particles are particles that cannot be distinguished from one another, even in principle. Species of Identical particles include elementary particles such as electrons, as well as composite microscopic particles such as atoms and molecules.
	There are two main categories of Identical particles: bosons, which can share quantum states, and fermions, which are forbidden from sharing quantum states (this property of fermions is known as the Pauli exclusion principle.)
Mass	Mass is a concept used in the physical sciences to explain a number of observable behaviors, and in everyday usage, it is common to identify Mass with those resulting behaviors. In particular, Mass is commonly identified with weight. But according to our modern scientific understanding, the weight of an object results from the interaction of its Mass with a gravitational field, so while Mass is part of the explanation of weight, it is not the complete explanation.
Motion	In physics, Motion means a change in the location of a body. Change in Motion is the result of applied force. Motion is typically described in terms of velocity, acceleration, displacement, and time.
Reduced mass	Reduced mass is the 'effective' inertial mass appearing in the two-body problem of Newtonian mechanics. This is a quantity with the unit of mass, which allows the two-body problem to be solved as if it were a one-body problem. Note however that the mass determining the gravitational force is not reduced.
Angular momentum	In physics, Angular momentum is a measure of the motion of mass about a center of rotation; by analogy with momentum, the product of inertia (mass) and velocity, Angular momentum can be interpreted as the product of a rotational inertia (moment of inertia or angular mass) and angular velocity.
	In two dimensions, the Angular momentum of mass m with respect to a chosen origin is given by: $$L = mvr\sin\theta$$ where m is the mass, v is the speed (magnitude of the velocity vector), r is the distance from the origin and θ is the angle between the velocity and the radius vector. This equation can be interpreted as the product of the moment of inertia mr^2 times angular velocity vsin(θ) / r.
Binding energy	Binding energy is the mechanical energy required to disassemble a whole into separate parts. A bound system has typically a lower potential energy than its constituent parts; this is what keeps the system together. The usual convention is that this corresponds to a positive Binding energy.
Bosons	In particle physics, Bosons are particles which obey Bose-Einstein statistics; they are named after Satyendra Nath Bose and Albert Einstein. In contrast to fermions, which obey Fermi-Dirac statistics, several Bosons can occupy the same quantum state. Thus, Bosons with the same energy can occupy the same place in space.
Fermion	In particle physics, Fermion s are particles which obey Fermi-Dirac statistics; they are named after Enrico Fermi. In contrast to bosons, which have Bose-Einstein statistics, only one Fermion can occupy a quantum state at a given time; this is the Pauli Exclusion Principle. Thus, if more than one Fermion occupies the same place in space, the properties of each Fermion (e.g. its spin) must be different from the rest.

Hydrogen	Hydrogen is the chemical element with atomic number 1. It is represented by the symbol H. At standard temperature and pressure, Hydrogen is a colorless, odorless, nonmetallic, tasteless, highly flammable diatomic gas with the molecular formula H_2. With an atomic weight of 1.007 94 u, Hydrogen is the lightest element.
Positronium	Positronium is a system consisting of an electron and its anti-particle, a positron, bound together into an 'exotic atom'. The orbit of the two particles and the set of energy levels is similar to that of the hydrogen atom (electron and proton.) However, because of the reduced mass, the frequencies associated with the spectral lines are less than half of those of the corresponding hydrogen lines.
Pauli exclusion principle	The Pauli exclusion principle is a quantum mechanical principle formulated by Wolfgang Pauli in 1925. It states that no two identical fermions may occupy the same quantum state simultaneously. A more rigorous statement of this principle is that, for two identical fermions, the total wave function is anti-symmetric.
Stark effect	The Stark effect is the shifting and splitting of spectral lines of atoms and molecules due to the presence of an external static electric field. The amount of splitting and or shifting is called the Stark splitting or Stark shift. In general one distinguishes first- and second-order Stark effect s.
Spin	In quantum mechanics, Spin is a fundamental property of atomic nuclei, hadrons, and elementary particles. For particles with non-zero Spin, Spin direction (also called Spin for short) is an important intrinsic degree of freedom. The head-on collision of a quark (the red ball) from one proton (the orange ball) with a gluon (the green ball) from another proton with opposite Spin, Spin is represented by the blue arrows circling the protons and the quark.
Statistics	Statistics is a mathematical science pertaining to the collection, analysis, interpretation or explanation, and presentation of data. Statisticians improve the quality of data with the design of experiments and survey sampling. Statistics also provides tools for prediction and forecasting using data and statistical models.
Electron	The Electron is a subatomic particle that carries a negative electric charge. It has no known substructure and is believed to be a point particle. An Electron has a mass that is approximately 1836 times less than that of the proton.
Exchange operator	The Exchange operator is a quantum mechanical operator used in the field of quantum chemistry. Specifically, it is a term found in the Fock operator. It is defined as:

$$\hat{K}_j(1)f(1) = \varphi_j(1) \int \frac{\varphi_j^*(2)f(2)}{r_{12}} dv_2$$

where $\hat{K}_j(1)$ is the one-electron Exchange operator,
f(1),f(2) are the one-electron wavefunction being acted upon by the Exchange operator as functions of the positions of electron 1 and electron 2,
$\varphi_j(1),\varphi_j(2)$ are the one-electron wavefunction of the j-th electron as functions of the positions of electron 1 and electron 2,
r_{12} is the distance between electrons 1 and 2.

Exchange force	In particle physics, an Exchange force is a force produced by the exchange of force carrier particles, such as the electromagnetic force produced by the exchange of photons between electrons and the strong force produced by the exchange of gluons between quarks. The idea of an Exchange force implies a continuous exchange of particles which accompany the interaction and transmit the force, a process that receives its operational justification through the Heisenberg uncertainty principle. These Exchange force s are not the same as the exchange interaction, also sometimes called the Exchange force between electrons which arises from a combination of the identity of particles, exchange symmetry, and the electrostatic force.
Boltzmann factor	In physics, the Boltzmann factor is a weighting factor that determines the relative probability of a state i in a multi-state system in thermodynamic equilibrium at temperature T. $$e^{-\frac{E_i}{k_B T}}$$ Where k_B is Boltzmann's constant, and E_i is the energy of state i. The ratio of the probabilities of two states is given by the ratio of their Boltzmann factor s.
Atom	The Atom is a basic unit of matter consisting of a dense, central nucleus surrounded by a cloud of negatively charged electrons. The Atom ic nucleus contains a mix of positively charged protons and electrically neutral neutrons (except in the case of hydrogen-1, which is the only stable nuclide with no neutron.) The electrons of an Atom are bound to the nucleus by the electromagnetic force.
Singlet	In theoretical physics, a singlet usually refers to a one-dimensional representation (e.g. a particle with vanishing spin.) It may also refer to two or more particles prepared in a correlated state, such that the total angular momentum of the state is zero. Singlets frequently occur in atomic physics as one of the two ways in which the spin of two electrons can be combined; the other being a triplet.
Hamiltonian	In quantum mechanics, the Hamiltonian H is the observable corresponding to the total energy of the system. It is a Hermitian matrix, that, when multiplied by the column vector representing the state of the system, gives a vector representing the total energy of the system. As with all observables, the spectrum of the Hamiltonian is the set of possible outcomes when one measures the total energy of a system.
Helium	Helium is the chemical element with atomic number 2, and is represented by the symbol He. It is a colorless, odorless, tasteless, non-toxic, inert monatomic gas that heads the noble gas group in the periodic table. Its boiling and melting points are the lowest among the elements and it exists only as a gas except in extreme conditions.
Helium atom	Helium is an element and the next simplest atom to solve after the hydrogen atom. Helium is composed of two electrons in orbit around a nucleus containing two protons along with some neutrons. The hydrogen atom is used extensively to aid in solving the Helium atom.
Screening	Screening is the damping of electric fields caused by the presence of mobile charge carriers. It is an important part of the behavior of charge-carrying fluids, such as ionized gases (classical plasmas) and conduction electrons in semiconductors and metals. In a fluid composed of electrically charged constituent particles, each pair of particles interact through the Coulomb force,

$$\mathbf{F} = \frac{q_1 q_2}{4\pi\epsilon_0 \left|\mathbf{r}\right|^2}\hat{\mathbf{r}}$$

Clebsch-Gordan coefficients	Clebsch-Gordan coefficients are the expansion coefficients of total angular momentum eigenstates in an uncoupled tensor product basis.
	Below, this definition is made precise by defining angular momentum operators, angular momentum eigenstates, and tensor products of these states.
	From the formal definition of angular momentum, recursion relations for the Clebsch-Gordan coefficients can be found.
Coefficient	In mathematics, a coefficient is a constant multiplicative factor of a specific object. For example, in the expression $9x^2$, the coefficient of x^2 is 9.
	The object can be such things as a variable, a vector, a function, etc.
Electron configuration	In atomic physics and quantum chemistry, Electron configuration is the arrangement of electrons of an atom, a molecule, or other physical structure. It concerns the way electrons can be distributed in the orbitals of the given system (atomic or molecular for instance.)
	Like other elementary particles, the electron is subject to the laws of quantum mechanics, and exhibits both particle-like and wave-like nature.
Gas	In physics, a Gas is a state of matter, consisting of a collection of particles (molecules, atoms, ions, electrons, etc.) without a definite shape or volume that are in more or less random motion.
	Due to the electronic nature of the aforementioned particles, a 'force field' is present throughout the space around them.
Solid	The Solid state of matter is characterized by a distinct structural rigidity and virtual resistance to deformation (i.e. changes of shape and/or volume.) Most Solid s have high values both of Young's modulus and of the shear modulus of elasticity. This contrasts with most liquids or fluids, which have a low shear modulus, and typically exhibit the capacity for macroscopic viscous flow.
Proximity effect	At the atomic level, when two atoms come into proximity, the highest energy orbitals of the atoms change substantially and the electrons on the two atoms reorganize. One way to probe a correlated state is through the Proximity effect. This phenomenon occurs when the correlations present in one degenerate system 'leak' into another one with which it is in chemical equilibrium.
Valence electrons	In science, Valence electrons are the outermost electrons of an atom, which are important in determining how the atom reacts chemically with other atoms. Atoms with a complete shell of Valence electrons tend to be chemically inert. Atoms with one or two Valence electrons more than a closed shell are highly reactive because the extra electrons are easily removed to form positive ions.

Fermi surface	In condensed matter physics, the Fermi surface is an abstract boundary useful for predicting the thermal, electrical, magnetic, and optical properties of metals, semimetals, and doped semiconductors. The shape of the Fermi surface is derived from the periodicity and symmetry of the crystalline lattice and from the occupation of electronic energy bands. The existence of a Fermi surface is a direct consequence of the Pauli exclusion principle, which allows a maximum of one electron per quantum state.
Density	The Density of a material is defined as its mass per unit volume. The symbol of Density is ρ '>rho.)
	Mathematically:
	$$\rho = \frac{m}{V}$$
	where:
	ρ is the Density,
	m is the mass,
	V is the volume.
Electron density	Electron density is the measure of the probability of an electron being present at a specific location.
	In molecules, regions of Electron density are usually found around the atom, and its bonds. In de-localized or conjugated systems, such as phenol, benzene and compounds such as hemoglobin and chlorophyll, the Electron density covers an entire region, i.e., in benzene they are found above and below the planar ring.
Thermal	A Thermal column (or Thermal is a column of rising air in the lower altitudes of the Earth's atmosphere. Thermal s are created by the uneven heating of the Earth's surface from solar radiation, and an example of convection. The Sun warms the ground, which in turn warms the air directly above it.
Balmer series	The Balmer series or Balmer lines in atomic physics, is the designation of one of a set of six different named series describing the spectral line emissions of the hydrogen atom. The Balmer series is calculated using the Balmer formula, an empirical equation discovered by Johann Balmer in 1885.
	The visible spectrum of light from hydrogen displays four wavelengths, 410 nm, 434 nm, 486 nm, and 656 nm, that reflect emissions of photons by electrons in excited states transitioning to the quantum level described by the principal quantum number n equals 2.
Displacement	In fluid mechanics, Displacement occurs when an object is immersed in a fluid, pushing it out of the way and taking its place. The volume of the fluid displaced can then be measured, as in the illustration, and from this the volume of the immersed object can be deduced (the volume of the immersed object will be exactly equal to the volume of the displaced fluid.)
	An object that sinks displaces an amount of fluid equal to the object's volume.
Alpha decay	Alpha decay is a type of radioactive decay in which an atomic nucleus emits an alpha particle (two protons and two neutrons bound together into a particle identical to a helium nucleus) and transforms (or 'decays') into an atom with a mass number 4 less and atomic number 2 less. For example:

(The second form is preferred because the first form appears electrically unbalanced. Fundamentally, the recoiling nucleus is very quickly stripped of the two extra electrons which give it an unbalanced charge.

Hilbert space	The mathematical concept of a Hilbert space generalizes the notion of Euclidean space. It extends the methods of vector algebra from the two-dimensional plane and three-dimensional space to infinite-dimensional spaces. In more formal terms, a Hilbert space is an inner product space -- an abstract vector space in which distances and angles can be measured -- which is 'complete', meaning that if a sequence of vectors is Cauchy, then it converges to some limit within the space.
Mechanics	Mechanics is the branch of physics concerned with the behaviour of physical bodies when subjected to forces or displacements, and the subsequent effect of the bodies on their environment. The discipline has its roots in several ancient civilizations During the early modern period, scientists such as Galileo, Kepler, and especially Newton, laid the foundation for what is now known as classical Mechanics.
Quantum	In physics, a Quantum is an indivisible entity of a quantity that has the same units as the Planck constant and is related to both energy and momentum of elementary particles of matter (called fermions) and of photons and other bosons. The word comes from the Latin 'quantus', for 'how much.' Behind this, one finds the fundamental notion that a physical property may be 'quantized', referred to as 'quantization'. This means that the magnitude can take on only certain discrete numerical values, rather than any value, at least within a range.
Statistical mechanics	Statistical mechanics is the application of probability theory, which includes mathematical tools for dealing with large populations, to the field of mechanics, which is concerned with the motion of particles or objects when subjected to a force. It provides a framework for relating the microscopic properties of individual atoms and molecules to the macroscopic or bulk properties of materials that can be observed in everyday life, therefore explaining thermodynamics as a natural result of statistics and mechanics at the microscopic level.
	It provides a molecular-level interpretation of thermodynamic quantities such as work, heat, free energy, and entropy, allowing the thermodynamic properties of bulk materials to be related to the spectroscopic data of individual molecules.
Ideal gas	An Ideal gas is a theoretical gas composed of a set of randomly-moving point particles that interact only through elastic collisions. The Ideal gas concept is useful because it obeys the Ideal gas law, a simplified equation of state, and is amenable to analysis under statistical mechanics.
	At normal ambient conditions such as standard temperature and pressure, most real gases behave qualitatively like an Ideal gas.
Boltzmann constant	The Boltzmann constant is the physical constant relating energy at the particle level with temperature observed at the bulk level. It is the gas constant R divided by the Avogadro constant N_A: $$k = \frac{R}{N_A}.$$ It has the same units as entropy. It is named after the Austrian physicist Ludwig Boltzmann.
Chemical potential	Chemical potential, symbolized by μ, is a quantity first described by the American engineer, chemist and mathematical physicist Josiah Williard Gibbs. He defined it as follows:

Gibbs noted also that for the purposes of this definition, any chemical element or combination of elements in given proportions may be considered a substance, whether capable or not of existing by itself as a homogeneous body. Chemical potential is also referred to as partial molar Gibbs energy (.

Condensation

Condensation is the change of the physical state of aggregation (or simply state) of matter from gaseous phase into liquid phase. When the transition happens from the gaseous phase into the solid phase directly, bypassing the liquid phase, the change is called deposition.

Condensation commonly occurs when a vapor is cooled to its dew point, but the dew point can also be reached through compression.

Photon

In physics, a Photon is an elementary particle, the quantum of the electromagnetic field and the basic 'unit' of light and all other forms of electromagnetic radiation. It is also the force carrier for the electromagnetic force. The effects of this force are easily observable at both the microscopic and macroscopic level, because the Photon has no rest mass; this allows for interactions at long distances.

Beta decay

In nuclear physics, Beta decay is a type of radioactive decay in which a beta particle (an electron or a positron) is emitted. In the case of electron emission, it is referred to as beta minus (β^-), while in the case of a positron emission as beta plus (β^+.) Kinetic energy of beta particles has continuous spectrum ranging from 0 to maximal available energy (Q), which depends on parent and daughter nuclear states participating in the decay.

Perturbation	Perturbation is a term used in astronomy to describe alterations to an object's orbit caused by gravitational interactions with bodies external to the system formed by the object and its parent body (for example, a star and its planet particularly by the gravitational fields of the gas giants. In April 1996, Jupiter's gravitational influence caused the period of Comet Hale-Bopp's orbit to decrease from 4,206 to 2,380 years.
Perturbation theory	In quantum mechanics, Perturbation theory is a set of approximation schemes directly related to mathematical perturbation for describing a complicated quantum system in terms of a simpler one. The idea is to start with a simple system for which a mathematical solution is known, and add an additional 'perturbing' Hamiltonian representing a weak disturbance to the system. If the disturbance is not too large, the various physical quantities associated with the perturbed system (e.g. its energy levels and eigenstates) can, from considerations of continuity, be expressed as 'corrections' to those of the simple system.
Order	Order in a crystal lattice is the arrangement of some property with respect to atomic positions. It arises in charge ordering, spin ordering, magnetic ordering, and compositional ordering. It is a thermodynamic entropy concept often displayed by a second Order phase transition.
Critical opalescence	Critical opalescence is a phenomenon which arises in the region of a continuous phase transition. Originally reported by Thomas Andrews in 1869 for the liquid-gas transition in carbon dioxide, many other examples have been discovered since. The phenomenon is most commonly demonstrated in binary fluid mixtures, such as methanol and cyclohexane.
Angle	In geometry and trigonometry, an Angle Where there is no possibility of confusion, the term Angle is used interchangeably for both the geometric configuration itself and for its angular magnitude (which is simply a numerical quantity.)
Degenerate	In physics two or more different physical states are said to be degenerate if they are all at the same energy level. Physical states differ if and only if they are linearly independent. An energy level is said to be degenerate if it contains two or more different states.
Balmer series	The Balmer series or Balmer lines in atomic physics, is the designation of one of a set of six different named series describing the spectral line emissions of the hydrogen atom. The Balmer series is calculated using the Balmer formula, an empirical equation discovered by Johann Balmer in 1885.
	The visible spectrum of light from hydrogen displays four wavelengths, 410 nm, 434 nm, 486 nm, and 656 nm, that reflect emissions of photons by electrons in excited states transitioning to the quantum level described by the principal quantum number n equals 2.
Fin	In the study of heat transfer, a Fin is a surface that extends from an object to increase the rate of heat transfer to or from the environment by increasing convection. The amount of conduction, convection, or radiation of an object determines the amount of heat it transfers. Increasing the temperature difference between the object and the environment, increasing the convection heat transfer coefficient, or increasing the surface area of the object increases the heat transfer.
Fine structure	In atomic physics, the Fine structure describes the splitting of the spectral lines of atoms due to first order relativistic corrections.

The gross structure of line spectra is the line spectra predicted by non-relativistic electrons with no spin. For a hydrogenic atom, the gross structure energy levels only depend on the principal quantum number n.

Hydrogen

Hydrogen is the chemical element with atomic number 1. It is represented by the symbol H. At standard temperature and pressure, Hydrogen is a colorless, odorless, nonmetallic, tasteless, highly flammable diatomic gas with the molecular formula H_2. With an atomic weight of 1.007 94 u, Hydrogen is the lightest element.

Hamiltonian

In quantum mechanics, the Hamiltonian H is the observable corresponding to the total energy of the system. It is a Hermitian matrix, that, when multiplied by the column vector representing the state of the system, gives a vector representing the total energy of the system. As with all observables, the spectrum of the Hamiltonian is the set of possible outcomes when one measures the total energy of a system.

Coupling

In electronics and telecommunication, Coupling is the desirable or undesirable transfer of energy from one medium, such as a metallic wire or an optical fiber, to another medium, including fortuitous transfer.

Coupling is also the transfer of power from one circuit segment to another, e.g., an alternating voltage may be transferred to a segment at a different direct voltage by use of a capacitor or transformer; power may be efficiently transferred to a segment with different impedance by use of a transformer.
Electromagnetic Coupling:

· inductive Coupling, most commonly transformer Coupling, also called magnetic Coupling
· capacitive Coupling, capacitor Coupling, also called electrostatic Coupling
· RF Coupling
· electromagnetic interference (EMI), sometimes called radio frequency interference (RFI), is unwanted Coupling. Electromagnetic compatibility (EMC) requires techniques to avoid such unwanted Coupling, such as electromagnetic shielding.
· wireless energy transfer
Other kinds of energy Coupling:

· acoustic Coupling with an acoustic coupler
· evanescent wave Coupling .

Energy

In physics, energy is a scalar physical quantity that describes the amount of work that can be performed by a force, an attribute of objects and systems that is subject to a conservation law. Different forms of energy include kinetic, potential, thermal, gravitational, sound, light, elastic, and electromagnetic energy The forms of energy are often named after a related force.

Kinetic energy

The Kinetic energy of an object is the extra energy which it possesses due to its motion. It is defined as the work needed to accelerate a body of a given mass from rest to its current velocity. Having gained this energy during its acceleration, the body maintains this Kinetic energy unless its speed changes.

Bohr radius	In the Bohr model of the structure of an atom, put forward by Niels Bohr in 1913, electrons orbit a central nucleus. The model says that the electrons orbit only at certain distances from the nucleus, depending on their energy. In the simplest atom, hydrogen, a single electron orbits, and the smallest possible orbit for the electron, that with the lowest energy, is most likely to be found at a distance from the nucleus called the Bohr radius.
Frequency	Frequency is the number of occurrences of a repeating event per unit time. It is also referred to as temporal Frequency. The period is the duration of one cycle in a repeating event, so the period is the reciprocal of the Frequency.
Quantum	In physics, a Quantum is an indivisible entity of a quantity that has the same units as the Planck constant and is related to both energy and momentum of elementary particles of matter (called fermions) and of photons and other bosons. The word comes from the Latin 'quantus', for 'how much.' Behind this, one finds the fundamental notion that a physical property may be 'quantized', referred to as 'quantization'. This means that the magnitude can take on only certain discrete numerical values, rather than any value, at least within a range.
Lyman series	In physics and chemistry, the Lyman series is the series of transitions and resulting ultraviolet emission lines of the hydrogen atom as an electron goes from n ≥ 2 to n = 1 (where n is the principal quantum number referring to the energy level of the electron.) The transitions are named sequentially by Greek letters: from n = 2 to n = 1 is called Lyman-alpha, 3 to 1 is Lyman-beta, 4 to 1 is Lyman-gamma, etc. The series is named after its discoverer, Theodore Lyman.
Magnetic field	Magnetic field s surround magnetic materials and electric currents and are detected by the force they exert on other magnetic materials and moving electric charges. The Magnetic field at a given point, is specified by both a direction and a magnitude (or strength); as such it is a vector field.
	For the physics of magnetic materials, see magnetism and magnet, and more specifically ferromagnetism, paramagnetism, and diamagnetism.
Dipole	In physics, there are two kinds of Dipole s :
	· An electric Dipole is a separation of positive and negative charge. The simplest example of this is a pair of electric charges of equal magnitude but opposite sign, separated by some, usually small, distance. A permanent electric Dipole is called an electret. · A magnetic Dipole is a closed circulation of electric current. A simple example of this is a single loop of wire with some constant current flowing through it.
	Dipole s can be characterized by their Dipole moment, a vector quantity. For the simple electric Dipole given above, the electric Dipole moment would point from the negative charge towards the positive charge, and have a magnitude equal to the strength of each charge times the separation between the charges. For the current loop, the magnetic Dipole moment would point through the loop (according to the right hand grip rule), with a magnitude equal to the current in the loop times the area of the loop.
	In addition to current loops, the electron, among other fundamental particles, is said to have a magnetic Dipole moment.
Thomas precession	In physics the Thomas precession is a special relativistic correction to the precession of a gyroscope in a rotating non-inertial frame. It can be understood as a consequence of the fact that the space of velocities in relativity is hyperbolic, and so parallel transport of a vector (the gyroscope's angular velocity) around a circle (its linear velocity) leaves it pointing in a different direction.

Thomas precession in relativity was already known to Ludwik Silberstein, in 1914.

Alpha decay	Alpha decay is a type of radioactive decay in which an atomic nucleus emits an alpha particle (two protons and two neutrons bound together into a particle identical to a helium nucleus) and transforms (or 'decays') into an atom with a mass number 4 less and atomic number 2 less. For example:
	(The second form is preferred because the first form appears electrically unbalanced. Fundamentally, the recoiling nucleus is very quickly stripped of the two extra electrons which give it an unbalanced charge.
Electron	The Electron is a subatomic particle that carries a negative electric charge. It has no known substructure and is believed to be a point particle. An Electron has a mass that is approximately 1836 times less than that of the proton.
G-factor	Where μ is the total magnetic moment resulting from both spin and orbital angular momentum of an electron, $J = L+S$ is its total angular momentum, and μ_B is the Bohr magneton
	Protons, neutrons, and many nuclei have spin and magnetic moments, and therefore associated G-factor s.
Magnetic moment	The Magnetic moment of a system is a measure of the strength and the direction of its magnetism. More technically (in physics, astronomy, chemistry, and electrical engineering), the term Magnetic moment of a system (such as a loop of electric current, a bar magnet, an electron, a molecule, or a planet) usually refers to its magnetic dipole moment, and quantifies the contribution of the system's internal magnetism to the external dipolar magnetic field produced by the system (that is, the component of the external magnetic field that drops off with distance as the inverse cube.) Any dipolar magnetic field pattern is symmetric with respect to rotations around a particular axis, therefore it is customary to describe the magnetic dipole moment that creates such a field as a vector with a direction along that axis.
Quantum electrodynamics	Quantum electrodynamics (Quantum electrodynamics D) is a relativistic quantum field theory of electrodynamics. Quantum electrodynamics D was developed by a number of physicists, beginning in the late 1920s. It basically describes how light and matter interact.
Spin-orbit interaction	In quantum physics, the Spin-orbit interaction is any interaction of a particle's spin with its motion. The first and best known example of this is that Spin-orbit interaction causes shifts in an electron's atomic energy levels, due to electromagnetic interaction between the electron's spin and the nucleus's electric field, through which it moves. A similar effect, due to the relationship between angular momentum and the strong nuclear force, occurs for protons and neutrons moving inside the nucleus, leading to a shift in their energy levels in the nucleus shell model.
Atom	The Atom is a basic unit of matter consisting of a dense, central nucleus surrounded by a cloud of negatively charged electrons. The Atom ic nucleus contains a mix of positively charged protons and electrically neutral neutrons (except in the case of hydrogen-1, which is the only stable nuclide with no neutron.) The electrons of an Atom are bound to the nucleus by the electromagnetic force.
Hydrogen atom	A Hydrogen atom is an atom of the chemical element hydrogen, and an example of a Boson. The electrically neutral atom contains a single positively-charged proton and a single negatively-charged electron bound to the nucleus by the Coulomb force. The most abundant isotope, hydrogen-1, protium, or light hydrogen, contains no neutrons; other isotopes contain one or more neutrons.

Clebsch-Gordan coefficients	Clebsch-Gordan coefficients are the expansion coefficients of total angular momentum eigenstates in an uncoupled tensor product basis.
	Below, this definition is made precise by defining angular momentum operators, angular momentum eigenstates, and tensor products of these states.
	From the formal definition of angular momentum, recursion relations for the Clebsch-Gordan coefficients can be found.

Dirac equation

In physics, the Dirac equation is a relativistic quantum mechanical wave equation formulated by British physicist Paul Dirac in 1928 which provides a description of elementary 1/2 href='/wiki/Spin-%C2%BD'>spin-1/2 particles, such as electrons, consistent with both the principles of quantum mechanics and the theory of special relativity. The equation demands the existence of antiparticles and actually predated their experimental discovery, making the discovery of the positron, the antiparticle of the electron, one of the greatest triumphs of modern theoretical physics.

The Dirac equation in the form originally proposed by Dirac is:

$$\left(\beta mc^2 + \sum_{k=1}^{3} \alpha_k p_k\, c \right) \psi(\mathbf{x},t) = i\hbar \frac{\partial \psi(\mathbf{x},t)}{\partial t}$$

where

m is the rest mass of the electron,
c is the speed of light,
p is the momentum operator,
$\hbar$ is the reduced Planck's constant,
x and t are the space and time coordinates.

Angular momentum

In physics, Angular momentum is a measure of the motion of mass about a center of rotation; by analogy with momentum, the product of inertia (mass) and velocity, Angular momentum can be interpreted as the product of a rotational inertia (moment of inertia or angular mass) and angular velocity.

In two dimensions, the Angular momentum of mass m with respect to a chosen origin is given by:

$$L = mvr \sin \theta$$

where m is the mass, v is the speed (magnitude of the velocity vector), r is the distance from the origin and θ is the angle between the velocity and the radius vector. This equation can be interpreted as the product of the moment of inertia mr^2 times angular velocity vsin(θ) / r.

Coefficient

In mathematics, a coefficient is a constant multiplicative factor of a specific object. For example, in the expression $9x^2$, the coefficient of x^2 is 9.

The object can be such things as a variable, a vector, a function, etc.

Bohr magneton

In atomic physics, the Bohr magneton is a physical constant of magnetic moment of electrons. It was discovered in 1913 by Romanian physicist Åžtefan Procopiu and rediscovered independently two years later by Danish physicist Niels Bohr. It is sometimes called the Bohr-Procopiu magneton.

Hyperfine structure	The term Hyperfine structure refers to a collection of different effects leading to small shifts and splittings in the energy levels of atoms, molecules and ions. The name is a reference to the fine structure which results from the interaction between the magnetic moments associated with the electron spin and orbital angular momentum. Hyperfine structure, with energy shifts typically orders of magnitude smaller than the fine structure, results from the interactions of the nucleus (or nuclei in molecules) with internally generated electric and magnetic fields.
Singlet	In theoretical physics, a singlet usually refers to a one-dimensional representation (e.g. a particle with vanishing spin.) It may also refer to two or more particles prepared in a correlated state, such that the total angular momentum of the state is zero.
	Singlets frequently occur in atomic physics as one of the two ways in which the spin of two electrons can be combined; the other being a triplet.
Exotic atom	An Exotic atom is an otherwise normal atom in which one or more sub-atomic particles have been replaced by other particles of the same charge. For example, electrons may be replaced by other negatively charged particles such as muons (muonic atoms) or pions (pionic atoms.) Because these substitute particles are usually unstable, Exotic atom s typically have short lifetimes.
Muonium	Muonium particles are exotic atoms made up of an antimuon and an electron, and are given the chemical symbol Mu. During the muon's 2 Åμs lifetime, Muonium can enter into compounds such as Muonium chloride (MuCl) or sodium muonide (NaMu.)
	Due to the mass difference between the antimuon and the electron, Muonium is more similar to atomic hydrogen than positronium.
Positronium	Positronium is a system consisting of an electron and its anti-particle, a positron, bound together into an 'exotic atom'. The orbit of the two particles and the set of energy levels is similar to that of the hydrogen atom (electron and proton.) However, because of the reduced mass, the frequencies associated with the spectral lines are less than half of those of the corresponding hydrogen lines.
Stark effect	The Stark effect is the shifting and splitting of spectral lines of atoms and molecules due to the presence of an external static electric field. The amount of splitting and or shifting is called the Stark splitting or Stark shift. In general one distinguishes first- and second-order Stark effect s.
Electric dipole moment	In physics, the Electric dipole moment is a measure of the separation of positive and negative electrical charges in a system of charges, that is, a measure of the charge system's overall polarity.
	In the simple case of two point charges, one with charge + q and one with charge − q, the Electric dipole moment p is: $$\boldsymbol{p} = q\,\boldsymbol{d}$$ where d is the displacement vector pointing from the negative charge to the positive charge. Thus, the Electric dipole moment vector p points from the negative charge to the positive charge.

Alpha decay	Alpha decay is a type of radioactive decay in which an atomic nucleus emits an alpha particle (two protons and two neutrons bound together into a particle identical to a helium nucleus) and transforms (or 'decays') into an atom with a mass number 4 less and atomic number 2 less. For example: (The second form is preferred because the first form appears electrically unbalanced. Fundamentally, the recoiling nucleus is very quickly stripped of the two extra electrons which give it an unbalanced charge.
Stark effect	The Stark effect is the shifting and splitting of spectral lines of atoms and molecules due to the presence of an external static electric field. The amount of splitting and or shifting is called the Stark splitting or Stark shift. In general one distinguishes first- and second-order Stark effect s.
Satellite Test of the Equivalence Principle	The Satellite Test of the Equivalence Principle is a proposed (as of 2008) space science experiment to test the equivalence principle of general relativity. The experiment is thought to be sensitive enough to test Einstein's theory of gravity and other theories. The basic configuration is that of a drag-free satellite where an outer shell around an inner test mass is used to block solar wind, atmospheric drag, the Earth's magnetic field and other effects which might disturb the motion of a freely-falling inner object.
Helium	Helium is the chemical element with atomic number 2, and is represented by the symbol He. It is a colorless, odorless, tasteless, non-toxic, inert monatomic gas that heads the noble gas group in the periodic table. Its boiling and melting points are the lowest among the elements and it exists only as a gas except in extreme conditions.
Ionization	Ionization is the physical process of converting an atom or molecule into an ion by adding or removing charged particles such as electrons or other ions. This is often confused with dissociation (chemistry.) The process works slightly differently depending on whether an ion with a positive or a negative electric charge is being produced.
Perturbation	Perturbation is a term used in astronomy to describe alterations to an object's orbit caused by gravitational interactions with bodies external to the system formed by the object and its parent body (for example, a star and its planet particularly by the gravitational fields of the gas giants. In April 1996, Jupiter's gravitational influence caused the period of Comet Hale-Bopp's orbit to decrease from 4,206 to 2,380 years.
Perturbation theory	In quantum mechanics, Perturbation theory is a set of approximation schemes directly related to mathematical perturbation for describing a complicated quantum system in terms of a simpler one. The idea is to start with a simple system for which a mathematical solution is known, and add an additional 'perturbing' Hamiltonian representing a weak disturbance to the system. If the disturbance is not too large, the various physical quantities associated with the perturbed system (e.g. its energy levels and eigenstates) can, from considerations of continuity, be expressed as 'corrections' to those of the simple system.
Critical opalescence	Critical opalescence is a phenomenon which arises in the region of a continuous phase transition. Originally reported by Thomas Andrews in 1869 for the liquid-gas transition in carbon dioxide, many other examples have been discovered since. The phenomenon is most commonly demonstrated in binary fluid mixtures, such as methanol and cyclohexane.

Hydrogen	Hydrogen is the chemical element with atomic number 1. It is represented by the symbol H. At standard temperature and pressure, Hydrogen is a colorless, odorless, nonmetallic, tasteless, highly flammable diatomic gas with the molecular formula H_2. With an atomic weight of 1.007 94 u, Hydrogen is the lightest element.
Lithium	Lithium is the chemical element with atomic number 3, and is represented by the symbol Li. It is a soft alkali metal with a silver-white color. Under standard conditions it is the lightest metal and the least dense solid element.
Atomic orbital	An Atomic orbital is a mathematical function that describes the wave-like behavior of either one electron or a pair of electrons, in an atom. This function can be used to calculate the probability of finding any electron of an atom in any specific region around the atom's nucleus. These functions may serve as three-dimensional graph of an electron's likely location.
Quantum	In physics, a Quantum is an indivisible entity of a quantity that has the same units as the Planck constant and is related to both energy and momentum of elementary particles of matter (called fermions) and of photons and other bosons. The word comes from the Latin 'quantus', for 'how much.' Behind this, one finds the fundamental notion that a physical property may be 'quantized', referred to as 'quantization'. This means that the magnitude can take on only certain discrete numerical values, rather than any value, at least within a range.

Potential well	A Potential well is the region surrounding a local minimum of potential energy. Energy captured in a Potential well is unable to convert to another type of energy (kinetic energy in the case of a gravitational Potential well) because it is captured in the local minimum of a Potential well. Therefore, a body may not proceed to the global minimum of potential energy, as it would naturally tend to due to entropy.
Coulomb barrier	The Coulomb barrier which is named after physicist Charles-Augustin de Coulomb (1736-1806), is the energy barrier due to electrostatic interaction that two nuclei need to overcome so they can get close enough to undergo nuclear fusion. This energy barrier is given by the electrostatic potential energy: $$U_{coul} = k\frac{q_1\,q_2}{r} = \frac{1}{4\pi\epsilon_0}\frac{q_1\,q_2}{r}$$ where
	k is the Coulomb's constant = 8.9876×10^9 N m^2 C^{-2}; ϵ_0 is the permittivity of free space; q_1, q_2 are the charges of the interacting particles; r is the interaction radius.
	A positive value of U is due to a repulsive force, so interacting particles are at higher energy levels as they get closer.
Alpha decay	Alpha decay is a type of radioactive decay in which an atomic nucleus emits an alpha particle (two protons and two neutrons bound together into a particle identical to a helium nucleus) and transforms (or 'decays') into an atom with a mass number 4 less and atomic number 2 less. For example:
	(The second form is preferred because the first form appears electrically unbalanced. Fundamentally, the recoiling nucleus is very quickly stripped of the two extra electrons which give it an unbalanced charge.
Clebsch-Gordan coefficients	Clebsch-Gordan coefficients are the expansion coefficients of total angular momentum eigenstates in an uncoupled tensor product basis.
	Below, this definition is made precise by defining angular momentum operators, angular momentum eigenstates, and tensor products of these states.
	From the formal definition of angular momentum, recursion relations for the Clebsch-Gordan coefficients can be found.
Coefficient	In mathematics, a coefficient is a constant multiplicative factor of a specific object. For example, in the expression $9x^2$, the coefficient of x^2 is 9.
	The object can be such things as a variable, a vector, a function, etc.
Angle	In geometry and trigonometry, an Angle Where there is no possibility of confusion, the term Angle is used interchangeably for both the geometric configuration itself and for its angular magnitude (which is simply a numerical quantity.)
Harmonic	In acoustics and telecommunication, a Harmonic of a wave is a component frequency of the signal that is an integer multiple of the fundamental frequency. For example, if the fundamental frequency is f, the Harmonic s have frequencies f, 2f, 3f, 4f, etc. The Harmonic s have the property that they are all periodic at the fundamental frequency, therefore the sum of Harmonic s is also periodic at that frequency.

Harmonic oscillator	In classical mechanics, a Harmonic oscillator is a system which, when displaced from its equilibrium position, experiences a restoring force, F, proportional to the displacement, x according to Hooke's law: $$F = -kx$$ where k is a positive constant.
	If F is the only force acting on the system, the system is called a simple Harmonic oscillator, and it undergoes simple harmonic motion: sinusoidal oscillations about the equilibrium point, with a constant amplitude and a constant frequency (which does not depend on the amplitude.)
	If a frictional force (damping) proportional to the velocity is also present, the Harmonic oscillator is described as a damped oscillator.
Atom	The Atom is a basic unit of matter consisting of a dense, central nucleus surrounded by a cloud of negatively charged electrons. The Atom ic nucleus contains a mix of positively charged protons and electrically neutral neutrons (except in the case of hydrogen-1, which is the only stable nuclide with no neutron.) The electrons of an Atom are bound to the nucleus by the electromagnetic force.
Hydrogen	Hydrogen is the chemical element with atomic number 1. It is represented by the symbol H. At standard temperature and pressure, Hydrogen is a colorless, odorless, nonmetallic, tasteless, highly flammable diatomic gas with the molecular formula H_2. With an atomic weight of 1.007 94 u, Hydrogen is the lightest element.
Hydrogen atom	A Hydrogen atom is an atom of the chemical element hydrogen, and an example of a Boson. The electrically neutral atom contains a single positively-charged proton and a single negatively-charged electron bound to the nucleus by the Coulomb force. The most abundant isotope, hydrogen-1, protium, or light hydrogen, contains no neutrons; other isotopes contain one or more neutrons.
Stark effect	The Stark effect is the shifting and splitting of spectral lines of atoms and molecules due to the presence of an external static electric field. The amount of splitting and or shifting is called the Stark splitting or Stark shift. In general one distinguishes first- and second-order Stark effect s.

Thomas precession	In physics the Thomas precession is a special relativistic correction to the precession of a gyroscope in a rotating non-inertial frame. It can be understood as a consequence of the fact that the space of velocities in relativity is hyperbolic, and so parallel transport of a vector (the gyroscope's angular velocity) around a circle (its linear velocity) leaves it pointing in a different direction. Thomas precession in relativity was already known to Ludwik Silberstein, in 1914.
Dynamics	In physics the term Dynamics customarily refers to the time evolution of physical processes. These processes may be microscopic as in particle physics, kinetic theory, and chemical reactions, or macroscopic as in the predictions of statistical mechanics and nonequilibrium thermodynamics.
Perturbation	Perturbation is a term used in astronomy to describe alterations to an object's orbit caused by gravitational interactions with bodies external to the system formed by the object and its parent body (for example, a star and its planet particularly by the gravitational fields of the gas giants. In April 1996, Jupiter's gravitational influence caused the period of Comet Hale-Bopp's orbit to decrease from 4,206 to 2,380 years.
Perturbation theory	In quantum mechanics, Perturbation theory is a set of approximation schemes directly related to mathematical perturbation for describing a complicated quantum system in terms of a simpler one. The idea is to start with a simple system for which a mathematical solution is known, and add an additional 'perturbing' Hamiltonian representing a weak disturbance to the system. If the disturbance is not too large, the various physical quantities associated with the perturbed system (e.g. its energy levels and eigenstates) can, from considerations of continuity, be expressed as 'corrections' to those of the simple system.
Quantum	In physics, a Quantum is an indivisible entity of a quantity that has the same units as the Planck constant and is related to both energy and momentum of elementary particles of matter (called fermions) and of photons and other bosons. The word comes from the Latin 'quantus', for 'how much.' Behind this, one finds the fundamental notion that a physical property may be 'quantized', referred to as 'quantization'. This means that the magnitude can take on only certain discrete numerical values, rather than any value, at least within a range.
Critical opalescence	Critical opalescence is a phenomenon which arises in the region of a continuous phase transition. Originally reported by Thomas Andrews in 1869 for the liquid-gas transition in carbon dioxide, many other examples have been discovered since. The phenomenon is most commonly demonstrated in binary fluid mixtures, such as methanol and cyclohexane.
Frequency	Frequency is the number of occurrences of a repeating event per unit time. It is also referred to as temporal Frequency. The period is the duration of one cycle in a repeating event, so the period is the reciprocal of the Frequency.
Absorption	In physics, Absorption of electromagnetic radiation is the way by which the energy of a photon is taken up by matter, typically the electrons of an atom. Thus, the electromagnetic energy is transformed to other forms of energy, for example, to heat. The Absorption of light during wave propagation is often called attenuation.
Emission	In physics, Emission is the process by which the energy of a photon is released by another entity, for example, by an atom whose electrons make a transition between two electronic energy levels. The emitted energy is in the form of a photon. The emittance of an object quantifies how much light is emitted by it.
Polarization	Polarization is a property of waves that describes the orientation of their oscillations

	By convention, the Polarization of light is described by specifying the direction of the wave's electric field.
Lyman series	In physics and chemistry, the Lyman series is the series of transitions and resulting ultraviolet emission lines of the hydrogen atom as an electron goes from n ≥ 2 to n = 1 (where n is the principal quantum number referring to the energy level of the electron.) The transitions are named sequentially by Greek letters: from n = 2 to n = 1 is called Lyman-alpha, 3 to 1 is Lyman-beta, 4 to 1 is Lyman-gamma, etc. The series is named after its discoverer, Theodore Lyman.
Dipole	In physics, there are two kinds of Dipole s :
	· An electric Dipole is a separation of positive and negative charge. The simplest example of this is a pair of electric charges of equal magnitude but opposite sign, separated by some, usually small, distance. A permanent electric Dipole is called an electret.
	· A magnetic Dipole is a closed circulation of electric current. A simple example of this is a single loop of wire with some constant current flowing through it.
	Dipole s can be characterized by their Dipole moment, a vector quantity. For the simple electric Dipole given above, the electric Dipole moment would point from the negative charge towards the positive charge, and have a magnitude equal to the strength of each charge times the separation between the charges. For the current loop, the magnetic Dipole moment would point through the loop (according to the right hand grip rule), with a magnitude equal to the current in the loop times the area of the loop.
	In addition to current loops, the electron, among other fundamental particles, is said to have a magnetic Dipole moment.
Electric dipole moment	In physics, the Electric dipole moment is a measure of the separation of positive and negative electrical charges in a system of charges, that is, a measure of the charge system's overall polarity.
	In the simple case of two point charges, one with charge + q and one with charge − q, the Electric dipole moment p is:
	$$p = q\,d$$
	where d is the displacement vector pointing from the negative charge to the positive charge. Thus, the Electric dipole moment vector p points from the negative charge to the positive charge.
Electromagnetic field	The Electromagnetic field is a physical field produced by electrically charged objects. It affects the behavior of charged objects in the vicinity of the field.
	The Electromagnetic field extends indefinitely throughout space and describes the electromagnetic interaction.
Photon	In physics, a Photon is an elementary particle, the quantum of the electromagnetic field and the basic 'unit' of light and all other forms of electromagnetic radiation. It is also the force carrier for the electromagnetic force. The effects of this force are easily observable at both the microscopic and macroscopic level, because the Photon has no rest mass; this allows for interactions at long distances.
Quantum electrodynamics	Quantum electrodynamics (Quantum electrodynamics D) is a relativistic quantum field theory of electrodynamics. Quantum electrodynamics D was developed by a number of physicists, beginning in the late 1920s. It basically describes how light and matter interact.

Spontaneous emission	Spontaneous emission is the process by which a light source such as an atom, molecule, nanocrystal or nucleus in an excited state undergoes a transition to the ground state and emits a photon. Spontaneous emission of light or luminescence is a fundamental process that plays an essential role in many phenomena in nature and forms the basis of many applications, such as fluorescent tubes, older television screens (cathode ray tubes), plasma display panels, lasers (for startup - normal continuous operation works by stimulated emission instead) and light emitting diodes.
	If a light source ('the atom') is in the excited state with energy E_2, it may spontaneously decay to the ground state, with energy E_1, releasing the difference in energy between the two states as a photon.
Stimulated emission	In optics, Stimulated emission is the process by which an electron, perturbed by a photon having the correct energy, may drop to a lower energy level resulting in the creation of another photon. The perturbing photon is seemingly unchanged in the process (cf. absorption), and the second photon is created with the same phase, frequency, polarization, and direction of travel as the original.
Larmor precession	In physics, Larmor precession is the precession of the magnetic moments of electrons, atomic nuclei, and atoms about an external magnetic field. The magnetic field exerts a torque on the magnetic moment,
	$$\vec{\Gamma} = \vec{\mu} \times \vec{B} = \gamma \vec{J} \times \vec{B}$$
	where $\vec{\Gamma}$ is the torque, $\vec{J}$ is the angular momentum vector, $\vec{B}$ is the external magnetic field, $\times$ is the cross product, and γ is the gyromagnetic ratio which gives the proportionality constant between the magnetic moment and the angular momentum. The angular momentum vector $\vec{J}$ precesses about the external field axis with an angular frequency known as the Larmor frequency,
	$\omega = \gamma B$
	$$\gamma = \frac{-eg}{2m}$$
	where ω is the angular frequency, $\gamma = \frac{-eg}{2m}$ is the gyromagnetic ratio, and B is the magnitude of the magnetic field and g is the g-factor (normally 1, except for in quantum physics.)
Boltzmann factor	In physics, the Boltzmann factor is a weighting factor that determines the relative probability of a state i in a multi-state system in thermodynamic equilibrium at temperature T.
	$$e^{-\frac{E_i}{k_B T}}$$
	Where k_B is Boltzmann's constant, and E_i is the energy of state i. The ratio of the probabilities of two states is given by the ratio of their Boltzmann factor s.
Coefficient	In mathematics, a coefficient is a constant multiplicative factor of a specific object. For example, in the expression $9x^2$, the coefficient of x^2 is 9.
	The object can be such things as a variable, a vector, a function, etc.
Excited state	Excitation is an elevation in energy level above an arbitrary baseline energy state. In physics there is a specific technical definition for energy level which is often associated with an atom being excited to an Excited state.

In quantum mechanics an Excited state of a system (such as an atom, molecule or nucleus) is any quantum state of the system that has a higher energy than the ground state (that is, more energy than the absolute minimum.)

Larmor formula	In physics, in the area of electrodynamics, the Larmor formula is used to calculate the total power radiated by a nonrelativistic point charge as it accelerates. It was first derived by J. J. Larmor in 1897, in the context of the wave theory of light.
	When accelerating or decelerating, any charged particle (such as an electron) radiates away energy in the form of electromagnetic waves.
Half-life	The Half-life of a quantity whose value decreases with time is the interval required for the quantity to decay to half of its initial value. The concept originated in describing how long it takes atoms to undergo radioactive decay but also applies in a wide variety of other situations.
	The term 'Half-life' dates to 1907.
Clebsch-Gordan coefficients	Clebsch-Gordan coefficients are the expansion coefficients of total angular momentum eigenstates in an uncoupled tensor product basis.
	Below, this definition is made precise by defining angular momentum operators, angular momentum eigenstates, and tensor products of these states.
	From the formal definition of angular momentum, recursion relations for the Clebsch-Gordan coefficients can be found.
Angular momentum	In physics, Angular momentum is a measure of the motion of mass about a center of rotation; by analogy with momentum, the product of inertia (mass) and velocity, Angular momentum can be interpreted as the product of a rotational inertia (moment of inertia or angular mass) and angular velocity.
	In two dimensions, the Angular momentum of mass m with respect to a chosen origin is given by:

$$L = mvr \sin\theta$$

where m is the mass, v is the speed (magnitude of the velocity vector), r is the distance from the origin and θ is the angle between the velocity and the radius vector. This equation can be interpreted as the product of the moment of inertia mr^2 times angular velocity $v\sin(\theta) / r$.

Resonance	In physics, Resonance is the tendency of a system to oscillate at larger amplitude at some frequencies than at others. These are known as the system's Resonance frequencies (or resonant frequencies.) At a resonant frequency the frequency of oscillation does not change with changing amplitude.
Nuclear magnetic resonance	Nuclear magnetic resonance is a property that magnetic nuclei have in a magnetic field and applied electromagnetic (EM) pulse, which cause the nuclei to absorb energy from the EM pulse and radiate this energy back out. The energy radiated back out is at a specific resonance frequency which depends on the strength of the magnetic field and other factors. This allows the observation of specific quantum mechanical magnetic properties of an atomic nucleus.
Quadrupole	A Quadrupole or quadrapole is one of a sequence of configurations of -- for example -- electric charge or current but it is usually just part of a multipole expansion of a more complex structure reflecting various orders of complexity.

The traceless Quadrupole moment tensor of a system of charges (or masses, for example) is defined as

$$Q_{ij} = \sum_n q_n (3x_i x_j - r^2 \delta_{ij}) \ ,$$

for a discrete system with individual charges q_n, or

$$Q_{ij} = \int \rho(x)(3x_i x_j - r^2 \delta_{ij}) \, d^3 x \ ,$$

for a continuous system with charge density $\rho(x)$.

The Quadrupole moment has 9 components, but because of the rotational symmetry and trace property, only 5 of these are independent.

Adiabatic theorem	The Adiabatic theorem is an important concept in quantum mechanics. Its original form, due to Max Born and Vladimir Fock (1928), can be stated as follows:
	A physical system remains in its instantaneous eigenstate if a given perturbation is acting on it slowly enough and if there is a gap between the eigenvalue and the rest of the Hamiltonian's spectrum.
	It may not be immediately clear from this formulation but the Adiabatic theorem is, in fact, an extremely intuitive concept.
Born-Oppenheimer	In quantum chemistry, the computation of the energy and wavefunction of an average-size molecule is a formidable task that is alleviated by the Born-Oppenheimer approximation. For instance the benzene molecule consists of 12 nuclei and 42 electrons. The time independent Schrödinger equation, which must be solved to obtain the energy and molecular wavefunction of this molecule, is a partial differential eigenvalue equation in 162 variables--the spatial coordinates of the electrons and the nuclei.
Pendulum	Derivations from Pendulum Figure 2. Force diagram of a simple gravity Pendulum.
	To begin, we shall make three assumptions about the simple Pendulum:
	· The rod/string/cable on which the bob is swinging is massless, does not stretch, and always remains taut. · The bob is a point mass. · Motion occurs in a 2-dimensional plane, i.e. Pendulum does not swing into and out of the page. Consider Figure 2. Note that the path of the Pendulum sweeps out an arc of a circle.
Angle	In geometry and trigonometry, an Angle Where there is no possibility of confusion, the term Angle is used interchangeably for both the geometric configuration itself and for its angular magnitude (which is simply a numerical quantity.)
Alpha decay	Alpha decay is a type of radioactive decay in which an atomic nucleus emits an alpha particle (two protons and two neutrons bound together into a particle identical to a helium nucleus) and transforms (or 'decays') into an atom with a mass number 4 less and atomic number 2 less. For example:
	(The second form is preferred because the first form appears electrically unbalanced. Fundamentally, the recoiling nucleus is very quickly stripped of the two extra electrons which give it an unbalanced charge.
Geometric phase	In mechanics (including classical mechanics as well as quantum mechanics), the Geometric phase, or the Pancharatnam-Berry phase , also known as the Pancharatnam phase or Berry phase, is a phase acquired over the course of a cycle, when the system is subjected to cyclic adiabatic processes, resulting from the geometrical properties of the parameter space of the Hamiltonian. The phenomenon was first discovered in 1956, and rediscovered in 1984. It can be seen in the Aharonov-Bohm effect and in the conical intersection of potential energy surfaces.
Lyman series	In physics and chemistry, the Lyman series is the series of transitions and resulting ultraviolet emission lines of the hydrogen atom as an electron goes from $n \geq 2$ to $n = 1$ (where n is the principal quantum number referring to the energy level of the electron.) The transitions are named sequentially by Greek letters: from $n = 2$ to $n = 1$ is called Lyman-alpha, 3 to 1 is Lyman-beta, 4 to 1 is Lyman-gamma, etc. The series is named after its discoverer, Theodore Lyman.

Magnetic flux	Magnetic flux, represented by the Greek letter Φ ' href='/wiki/Phi_(letter)'>phi), is a measure of quantity of magnetism, taking into account the strength and the extent of a magnetic field. The SI unit of Magnetic flux is the weber (in derived units: volt-seconds), and the unit of magnetic field is the weber per square meter, or tesla. Figure 1: The definition of surface integral relies on splitting the surface into small surface elements.
Rydberg formula	The Rydberg formula is used in atomic physics to describe the wavelengths of spectral lines of many chemical elements. The formula was invented by the Swedish physicist Johannes Rydberg and presented on November 5, 1888.
	In the 1880s, Rydberg worked on a formula describing the relation between the wavelengths in spectral lines of alkali metals.
Scalar potential	A Scalar potential is a fundamental concept in vector analysis and physics (the adjective 'scalar' is frequently omitted if there is no danger of confusion with vector potential.) Given a vector field F, its Scalar potential V is a scalar field whose negative gradient is F, $$\mathbf{F} = -\nabla V.$$ Conversely, given a function V, this formula defines a vector field F with the Scalar potential V. Scalar potential is also frequently denoted by the Greek letter Φ, for example, in electrodynamics.
Force	In physics, a Force is a push or pull that can cause an object with mass to change its velocity. Force has both magnitude and direction, making it a vector quantity. Newton's second law states that an object with a constant mass will accelerate in proportion to the net Force acting upon and in inverse proportion to its mass.
Balmer series	The Balmer series or Balmer lines in atomic physics, is the designation of one of a set of six different named series describing the spectral line emissions of the hydrogen atom. The Balmer series is calculated using the Balmer formula, an empirical equation discovered by Johann Balmer in 1885.
	The visible spectrum of light from hydrogen displays four wavelengths, 410 nm, 434 nm, 486 nm, and 656 nm, that reflect emissions of photons by electrons in excited states transitioning to the quantum level described by the principal quantum number n equals 2.
Flux quantization	Flux quantization is a quantum phenomenon in which the magnetic field is quantized in the unit of h / 2e, also known variously as flux quanta, fluxoids, vortices or fluxons.
	Flux quantization occurs in Type II superconductors subjected to a magnetic field. Below a critical field H_{c1}, all magnetic flux is expulsed according to the Meissner effect and perfect diamagnetism is observed, exactly as in a Type I superconductor.

Chapter 11. SCATTERING

Rydberg formula	The Rydberg formula is used in atomic physics to describe the wavelengths of spectral lines of many chemical elements. The formula was invented by the Swedish physicist Johannes Rydberg and presented on November 5, 1888.
	In the 1880s, Rydberg worked on a formula describing the relation between the wavelengths in spectral lines of alkali metals.
Angle	In geometry and trigonometry, an Angle Where there is no possibility of confusion, the term Angle is used interchangeably for both the geometric configuration itself and for its angular magnitude (which is simply a numerical quantity.)
Impact	In mechanics, an Impact is a high force or shock applied over a short time period. Such a force or acceleration can sometimes have a greater effect than a lower force applied over a proportionally longer time period.
	At normal speeds, during a perfectly inelastic collision, an object struck by a projectile will deform, and this deformation will absorb most, or even all, of the force of the collision.
Scattering	Scattering is a general physical process where some forms of radiation, such as light, sound are forced to deviate from a straight trajectory by one or more localized non-uniformities in the medium through which they pass. In conventional use, this also includes deviation of reflected radiation from the angle predicted by the law of reflection. Reflections that undergo Scattering are often called diffuse reflections and unscattered reflections are called specular (mirror-like) reflections.
Scattering theory	In mathematics and physics, Scattering theory is a framework for studying and understanding the scattering of waves and particles. Prosaically, wave scattering corresponds to the collision and scattering of a wave with some material object, for instance sunlight scattered by rain drops to form a rainbow. Scattering also includes the interaction of billiard balls on a table, the Rutherford scattering (or angle change) of alpha particles by gold nuclei, the Bragg scattering (or diffraction) of electrons and X-rays by a cluster of atoms, and the inelastic scattering of a fission fragment as it traverses a thin foil.
Scattering cross-section	The Scattering cross-section, σ_{scat}, relates the scattering of light or other radiation to the number of particles present. In terms of area, the total cross-section (σ) is the sum of the cross-sections due to absorption, scattering and luminescence: $\sigma = \sigma_A + \sigma_S + \sigma_L$
	The total cross-section is related to the absorbance of the light intensity through Beer-Lambert's law, which says absorbance is proportional to concentration: $A_\lambda = Cl\sigma$, where C is the concentration as a number density, A_λ is the absorbance at a given wavelength, λ, and l is the path length. The extinction or absorbance of the radiation is the logarithm (decadic or, more usually, natural) of the reciprocal of the transmittance: $A_\lambda = -\log T$
	There is no simple relationship between the Scattering cross-section and the physical size of the particles, as the Scattering cross-section depends on the wavelength of radiation used.
Rutherford	The Rutherford is an obsolete unit of radioactivity, defined as the activity of a quantity of radioactive material in which one million nuclei decay per second. It is therefore equivalent to one megabecquerel. It was named after Ernest Rutherford It is not an SI unit.

Rutherford scattering	In physics, Rutherford scattering is a phenomenon that was explained by Ernest Rutherford in 1909, and led to the development of the Rutherford model (planetary model) of the atom, and eventually to the Bohr model. It is now exploited by the materials analytical technique Rutherford backscattering. Rutherford scattering is also sometimes referred to as Coulomb scattering because it relies on static electric (Coulomb) forces.
Force	In physics, a Force is a push or pull that can cause an object with mass to change its velocity. Force has both magnitude and direction, making it a vector quantity. Newton's second law states that an object with a constant mass will accelerate in proportion to the net Force acting upon and in inverse proportion to its mass.
Frequency	Frequency is the number of occurrences of a repeating event per unit time. It is also referred to as temporal Frequency. The period is the duration of one cycle in a repeating event, so the period is the reciprocal of the Frequency.
Clebsch-Gordan coefficients	Clebsch-Gordan coefficients are the expansion coefficients of total angular momentum eigenstates in an uncoupled tensor product basis.
	Below, this definition is made precise by defining angular momentum operators, angular momentum eigenstates, and tensor products of these states.
	From the formal definition of angular momentum, recursion relations for the Clebsch-Gordan coefficients can be found.
Coefficient	In mathematics, a coefficient is a constant multiplicative factor of a specific object. For example, in the expression $9x^2$, the coefficient of x^2 is 9.
	The object can be such things as a variable, a vector, a function, etc.
Atom	The Atom is a basic unit of matter consisting of a dense, central nucleus surrounded by a cloud of negatively charged electrons. The Atom ic nucleus contains a mix of positively charged protons and electrically neutral neutrons (except in the case of hydrogen-1, which is the only stable nuclide with no neutron.) The electrons of an Atom are bound to the nucleus by the electromagnetic force.
Hydrogen	Hydrogen is the chemical element with atomic number 1. It is represented by the symbol H. At standard temperature and pressure, Hydrogen is a colorless, odorless, nonmetallic, tasteless, highly flammable diatomic gas with the molecular formula H_2. With an atomic weight of 1.007 94 u, Hydrogen is the lightest element.
Hydrogen atom	A Hydrogen atom is an atom of the chemical element hydrogen, and an example of a Boson. The electrically neutral atom contains a single positively-charged proton and a single negatively-charged electron bound to the nucleus by the Coulomb force. The most abundant isotope, hydrogen-1, protium, or light hydrogen, contains no neutrons; other isotopes contain one or more neutrons.
Energy	In physics, energy is a scalar physical quantity that describes the amount of work that can be performed by a force, an attribute of objects and systems that is subject to a conservation law. Different forms of energy include kinetic, potential, thermal, gravitational, sound, light, elastic, and electromagnetic energy The forms of energy are often named after a related force.

Chapter 11. SCATTERING

Low pressure area	A low pressure area is a region where the atmospheric pressure is lower in relation to the surrounding area. Low pressure systems form under areas of upper level divergence on the east side of upper troughs, or due to localized heating caused by greater insolation or active thunderstorm activity. Those that form due to organized thunderstorm activity over the water which acquire a well-defined circulation are called tropical cyclones.
Impulse	In classical mechanics, an Impulse is defined as the integral of a force with respect to time. When a force is applied to a rigid body it changes the momentum of that body. A small force applied for a long time can produce the same momentum change as a large force applied briefly, because it is the product of the force and the time for which it is applied that is important.
Propagator	In quantum mechanics and quantum field theory, the Propagator gives the probability amplitude for a particle to travel from one place to another in a given time, or to travel with a certain energy and momentum. Propagator s are used to represent the contribution of virtual particles on the internal lines of Feynman diagrams. They also can be viewed as the inverse of the wave operator appropriate to the particle, and are therefore often called Green's functions.
Hamiltonian	In quantum mechanics, the Hamiltonian H is the observable corresponding to the total energy of the system. It is a Hermitian matrix, that, when multiplied by the column vector representing the state of the system, gives a vector representing the total energy of the system. As with all observables, the spectrum of the Hamiltonian is the set of possible outcomes when one measures the total energy of a system.

Clebsch-Gordan coefficients	Clebsch-Gordan coefficients are the expansion coefficients of total angular momentum eigenstates in an uncoupled tensor product basis.
	Below, this definition is made precise by defining angular momentum operators, angular momentum eigenstates, and tensor products of these states.
	From the formal definition of angular momentum, recursion relations for the Clebsch-Gordan coefficients can be found.
Hamiltonian	In quantum mechanics, the Hamiltonian H is the observable corresponding to the total energy of the system. It is a Hermitian matrix, that, when multiplied by the column vector representing the state of the system, gives a vector representing the total energy of the system. As with all observables, the spectrum of the Hamiltonian is the set of possible outcomes when one measures the total energy of a system.
Coefficient	In mathematics, a coefficient is a constant multiplicative factor of a specific object. For example, in the expression $9x^2$, the coefficient of x^2 is 9.
	The object can be such things as a variable, a vector, a function, etc.
Complementarity	In physics, Complementarity is a basic principle of quantum theory closely identified with the Copenhagen interpretation, and refers to effects such as the wave-particle duality, in which different measurements made on a system reveal it to have either particle-like or wave-like properties. Niels Bohr is usually associated with this concept, which he developed at Copenhagen with Heisenberg, as a philosophical adjunct to the recently developed mathematics of quantum mechanics and in particular the Heisenberg uncertainty principle; in the narrow orthodox form, it is stated that a single quantum mechanical entity can either behave as a particle or as wave, but never simultaneously as both; that a stronger manifestation of the particle nature leads to a weaker manifestation of the wave nature and vice versa.
	The principle states that sometimes an object can have several (apparently) contradictory properties.
Duality	In electrical engineering, electrical terms are associated into pairs called duals. A dual of a relationship is formed by interchanging voltage and current in an expression. The dual expression thus produced is of the same form, and the reason that the dual is always a valid statement can be traced to the Duality of electricity and magnetism.
Wave-particle duality	In physics and chemistry, wave-particle duality is the concept that all matter and energy exhibits both wave-like and particle-like properties. A central concept of quantum mechanics, duality addresses the inadequacy of classical concepts like 'particle' and 'wave' in fully describing the behaviour of small-scale objects. Various interpretations of quantum mechanics attempt to explain this ostensible paradox.
David Joseph Bohm	David Joseph Bohm was an British quantum physicist who made significant contributions in the fields of theoretical physics, philosophy and neuropsychology, and to the Manhattan Project.
	Bohm was born in Wilkes-Barre, Pennsylvania to a Hungarian Jewish immigrant father and a Lithuanian Jewish mother. He was raised mainly by his father, a furniture store owner and assistant of the local rabbi.
Singlet	In theoretical physics, a singlet usually refers to a one-dimensional representation (e.g. a particle with vanishing spin.) It may also refer to two or more particles prepared in a correlated state, such that the total angular momentum of the state is zero.
	Singlets frequently occur in atomic physics as one of the two ways in which the spin of two electrons can be combined; the other being a triplet.

Chapter 12. AFTERWORD

John Stewart Bell	John Stewart Bell was a physicist, and the originator of Bell's Theorem, one of the most important theorems in quantum physics.
	He was born in Belfast, Northern Ireland, and graduated in experimental physics at the Queen's University of Belfast, in 1948. He went on to complete a PhD at the University of Birmingham, specialising in nuclear physics and quantum field theory.
Measurement	The framework of quantum mechanics requires a careful definition of measurement and a thorough discussion of its practical and philosophical implications.
	measurement is viewed in different ways in the many interpretations of quantum mechanics; however, despite the considerable philosophical differences, they almost universally agree on the practical question of what results from a routine quantum-physics laboratory measurement. To describe this, a simple framework to use is the Copenhagen interpretation, and it will be implicitly used in this section; the utility of this approach has been verified countless times, and all other interpretations are necessarily constructed so as to give the same quantitative predictions as this in almost every case.
Quantum	In physics, a Quantum is an indivisible entity of a quantity that has the same units as the Planck constant and is related to both energy and momentum of elementary particles of matter (called fermions) and of photons and other bosons. The word comes from the Latin 'quantus', for 'how much.' Behind this, one finds the fundamental notion that a physical property may be 'quantized', referred to as 'quantization'. This means that the magnitude can take on only certain discrete numerical values, rather than any value, at least within a range.
Rydberg formula	The Rydberg formula is used in atomic physics to describe the wavelengths of spectral lines of many chemical elements. The formula was invented by the Swedish physicist Johannes Rydberg and presented on November 5, 1888.
	In the 1880s, Rydberg worked on a formula describing the relation between the wavelengths in spectral lines of alkali metals.
Balmer series	The Balmer series or Balmer lines in atomic physics, is the designation of one of a set of six different named series describing the spectral line emissions of the hydrogen atom. The Balmer series is calculated using the Balmer formula, an empirical equation discovered by Johann Balmer in 1885.
	The visible spectrum of light from hydrogen displays four wavelengths, 410 nm, 434 nm, 486 nm, and 656 nm, that reflect emissions of photons by electrons in excited states transitioning to the quantum level described by the principal quantum number n equals 2.
Component	In thermodynamics, a Component is a chemically distinct constituent of a system. Calculating the number of components in a system is necessary, for example, when applying Gibbs phase rule in determination of the number of degrees of freedom of a system.
	The number of components is equal to the number of independent chemical constituents, minus the number of chemical reactions between them, minus the number of any constraints (like charge neutrality or balance of molar quantities.)
Degenerate	In physics two or more different physical states are said to be degenerate if they are all at the same energy level. Physical states differ if and only if they are linearly independent. An energy level is said to be degenerate if it contains two or more different states.

Taylor series | In mathematics, the Taylor series is a representation of a function as an infinite sum of terms calculated from the values of its derivatives at a single point. It may be regarded as the limit of the Taylor polynomials. Taylor series are named after the English mathematician Brook Taylor.

Lightning Source UK Ltd.
Milton Keynes UK
UKOW01f2301080514

231355UK00005B/55/P